IET ENERGY ENGINEERING SERIES 247

Power System Strength

Other volumes in this series:

Power System Strength

Evaluation methods, best practice,
case studies, and applications

Edited by
Hassan Haes Alhelou, Nasser Hosseinzadeh and
Behrooz Bahrani

The Institution of Engineering and Technology

Published by The Institution of Engineering and Technology, London, United Kingdom

The Institution of Engineering and Technology is registered as a Charity in England & Wales (no. 211014) and Scotland (no. SC038698).

First published 2023

The Institution of Engineering and Technology
Futures Place
Kings Way, Stevenage
Hertfordshire SG1 2UA, United Kingdom

www.theiet.org

British Library Cataloguing in Publication Data
A catalogue record for this product is available from the British Library

ISBN 978-1-83953-807-0 (hardback)
ISBN 978-1-83953-808-7 (PDF)

Typeset in India by MPS Limited

Cover Image: querbeet/E+ via Getty Images

Contents

About the editors

Hassan Haes Alhelou is with the Department of Electrical and Computer Systems Engineering of Monash University, Australia. He is a senior IEEE member, was listed in the Top 2% of scientists in the world by the Stanford University database in 2022, and the recipient of the Outstanding Reviewer Award from many journals, e.g., *Applied Energy*. He has participated in 15 international industrial projects globally. His research interests include power system dynamics and stability, operation and control, dynamic state estimation, and frequency control.

Nasser Hosseinzadeh is the director of the Centre for Smart Power and Energy Research of Deakin University, Australia. With a PhD from Victoria University, Australia, prior positions include head of the Department of Systems at CQ University Australia, and of the Department of Electrical and Computer Engineering, Sultan Qaboos University, Oman. His main research focus is stability assessment of the power grid as impacted by inverter-based resources (IBRs). He is particularly working on the impact of IBRs on the power system strength, and the impact of a weak system on the operation of IBRs. He serves as a reviewer for several IEEE journals.

Behrooz Bahrani is a senior lecturer at the Department of Electrical and Computer Systems Engineering at Monash University, Australia. He earned his PhD at the Ecole Polytechnique Federale de Lausanne (EPFL). He was a postdoctoral fellow at several institutions including EPFL, Purdue University, USA, Georgia Institute of Technology, USA, and Technical University of Munich, Germany. His research interests include control of power electronics systems, applications of power electronics in power and traction systems, and grid integration.

Foreword

For decades, traditional power systems were operated safely and securely with no problems related to their stability and security. This view of power systems has changed during the last decade due to environmental challenges and energy security risks which enforced the movement towards the feasible implementation of modern power systems based on the smart grid concept. In the last decade, the penetration level of renewable energy sources (RESs) besides inverter-based resources (IBRs) and inverter-based loads (IBLs) has been highly increased. Although RESs reduce environmental concerns but have negative impacts on the stability and security of existing power systems. Nowadays, one of the main concerns is about the system strength in power systems enriched with IBRs, IBLs, and RESs.

System strength is a crucial parameter in modern power systems besides system inertia since they both enable the system for maintaining an acceptable level of stability. System strength relates to the ability of a power system to manage fluctuations in supply or demand while maintaining stable voltage levels. Inertia relates to the ability of a power system to manage fluctuations in supply or demand while maintaining stable system frequency. In traditional power systems, the short circuit ratio (SCR) was used as an index for evaluating the system's strength. In fact, the SCR is the synchronous three-phase fault level (in MVA) divided by the rated output of an IBR generating system (in MW or MVA) measured at the generating system's connection point. The SCR is no more a valid index for evaluating the system's strength, especially in modern power systems. Power system operators around the globe, i.e., in USA, Australia, UK, and other modern systems, are looking for new metrics that can evaluate the strength of the system; hence, they can understand the hosting capacity of their modern power systems of renewables and that would accelerate the decarbonization plan by providing solutions for increasing the hosting capacity of green and renewable energy in modern systems.

In addition to the aforementioned challenge, there is a great research gap related to solutions for improving the system strength in light of the high penetrations of renewables. There are perspectives in providing these services from the demand side, especially by well-managing controllable loads, electric vehicles (EVs), IBRs, and IBLs. The promising field for providing such services is utilizing both the demand side and upstream by the deployment of active demand-side programs and grid-forming inverters for IBLs and IBRs.

The aforementioned description highlights the motivations for this comprehensive book on power system strength topic. This book covers the latest evaluation methods, best practices, case studies, and applications for encouraging researchers and governments for supporting the movement toward a smart grid concept of clean energy systems by providing solutions to low system strengths.

Introduction

The book is principally focused on power system strength and its importance for modern power systems. The book is sorted out and organized in ten chapters. Each chapter begins with the fundamental structure of the problem required for a rudimentary understanding of the methods described. The book starts with chapters that give comprehensive review and discussions on system strength and the challenges associated with it. The next set of chapters discusses the techniques designed for evaluating and assessing system strength followed by methods to improve the system strength in power systems. Brief descriptions of the book chapters are given below.

Chapter 1: The power system strength has been conventionally evaluated by finding short circuit ratio (SCR) at nodes of a power network. The short circuit index is still usually used although several other indices are defined for using in systems with inverter-based resources (IBRs). However, it has been found that these indices do not represent – to a good accuracy – the actual value of the system strength for a system with many IBRs and other power electronic devices such as static var compensators (SVCs) and other FACTS devices. There is obviously a pressing need for finding a new metric for correctly evaluating the system strength in modern power systems dominated by IBRs such as wind farms, solar power plants with storage, other renewable energy sources, microgrids, and active distribution networks. Although this can be done by detailed system modelling through EMT-based simulations, but there is a need for a metric that can evaluate system strength with a good approximation without a need for detailed modelling, similar to the way that SCR can be easily calculated. This chapter discusses the concept of power system strength in modern power grids in contrast with the system strength in conventional systems, and serves as an introduction to other chapters of this book as each of them will discuss some specific aspects of the power system strength notion.

Chapter 2: This chapter provides the theoretical formulations for defining power system strength followed by a comprehensive review of existing power system methodologies capturing the concepts, mathematical formulation, and assumptions along with research gaps. Several case studies to show the performance of short circuit ratio (SCR) index along with proposed solutions to mitigate system strength issues are presented.

Chapter 3: This chapter analyzes the power transfer limit (P_{max}) of inverter-based resources (IBRs) under the weak grid condition. It is pointed out that the impact of different grid parameters and grid configurations on P_{max} cannot be fully reflected by the short circuit ratio (SCR) and its extended forms, but can be readily captured by voltage sensitivity ($\partial Q/\partial V$). Hence, $\partial Q/\partial V$ turns out to be a more appropriate metric for the assessment of P_{max} compared to the SCR. The simulation results are given to demonstrate the theoretical analysis.

Chapter 4: The system strength assessment (SSA) is essential for contemporary power systems integrated with various invertor-based generators (IBGs). These IBGs intrinsically interfaced by power electronic converters and their intermittent energy sources such as wind and solar, create new type of challenges to power system operators and researchers. Moreover, ever increasing demands have overstretched the operating point (OP) of power systems, which are now operating closer to their instability points. Furthermore, the loads in the distribution system which were regarded as constant loads SSA, can no longer be considered as such with the advent of active distribution networks (ADNs). Concerning these factors, a swift and precise system strength assessment (SSA) is needed to operate the power system accurately without the loss of power to the customer. The SSA studies are typically conducted on a specific area of the power system, especially in the weak parts of the system, which is called the "study system" (SS). The area outside the SS is called the "external system" (ES), which shall be represented as a reduced order model in the SSA for accelerated results. With the evolution of power systems, the model reduction techniques branched into two fields, i.e., classical techniques and measurement-based techniques. Both techniques have their merits and drawbacks. Therefore, these two broad categories of power system reduction techniques are discussed in two chapters. In this chapter, the reduction techniques for dynamic equivalent models of the ES from the classical techniques are discussed. A case study is performed on simplified Australian 14 generator model to show the potential of classical techniques in the reduction of power systems, and their limitations in applying to modern IBRs rich power networks are examined.

Chapter 5: This chapter is an extension to Chapter 4. It introduces new techniques in reducing the size of modern large power systems including data-driven techniques, and measurement-based techniques. It also includes reduction techniques for wind power plants, solar power plants, microgrids, and active distribution networks. Also, a case study is presented on dynamic model reduction of a power system using LSTM recurrent neural networks. The chapter concludes that the best way of reducing a large power system model is a combination of classical and data-driven techniques. This is because in most cases, some data of the system is available for the model, and only a few dynamic data are unknown such as generator dynamics of only some generating units, for which data-driven techniques can be used to compute them approximately. In the parameter estimation process, a set of critical parameters should be selected based on sensitivity analysis to decrease the burden as well as reaching the local optimum. In this way, a robust model of the external system, i.e., all parts of the system except the area under study can be implemented for system strength assessment studies. A case study is performed on simplified Australian 14 generator model to show the potential of measurement-based techniques in the reduction of power systems and their strengths limitations and limitations in applying to modern IBRs-rich power networks are examined.

Chapter 6: This chapter provides an overview of the impact of integrating inverter-based resources (IBRs) on power system inertia and strength. Inertia is a crucial aspect in power systems for maintaining frequency stability, and the chapter explores the historical perspectives and importance of system inertia. In addition, with the increasing integration of IBRs, the chapter examines how they impact the power system's inertia and strength, highlighting the issues arising from the low inertia provided by IBRs and the reduction in the fault current contribution. Furthermore, the chapter delves into the frequency response and inertia requirements and presents different methods for estimating power system inertia. The case study of a power system with integrated wind energy plants illustrates the impact of IBRs on the power system's stability. Finally, the chapter discusses research gaps and industry challenges, as well as future research directions. The presented information provides valuable insights for power system operators and researchers to better manage the challenges posed by integrating IBRs into power systems.

Chapter 7: This chapter provides an overview of past ideas and perspectives and future visions of available transmission capacity (ATC). Methods can be classified into two general categories: dynamic and static, and the methods presented compete in both categories according to their accuracy and speed. This chapter introduces ATC, reviews conventional definitions, concepts, and terms, and, finally, evaluates the disadvantages and advantages of each method. Also, it is discussed why holomorphic algorithms are advantageous when determining ATC. ATC can also be determined faster and more accurately using "differential equation load flow." Due to the presence of HVDC lines and wind power plants, ATC calculations are challenging. The following sections explore the problems associated with inaccurate power network information and the use of state estimation in ATC calculations. Lastly, cyber-attacks to change ATC data are examined and their impact on the electricity market is discussed. The methods provided in this chapter will be very useful to future ATC researchers for developing faster and more accurate methods.

Chapter 8: In this chapter, an adaptive controller for a fifth-order grid-connected system with several other control objectives will be presented and implemented. The purpose of the proposed AFC is to keep track of the target $d–q$ current of the inverter where the uncertain dynamics of the PV grid-connected inverter system are estimated using a fuzzy logic approximator that uses the tracking error of the system. For the input–output feedback linearization control to be valid, the stability of the zero dynamics of the system is evaluated and proved to be stable.

Chapter 9: The transition to complete decarbonization is bringing about a total revolution in the electric power system paradigm. This chapter is focused on issues related to voltage stability and fault level decadence, proposing an effective procedure for short-circuit analysis including the latest generation of power electronic converters but always maintaining a steady-state approach. The computation tool is tested and exploited to perform studies on high-RES penetration grids in order to discuss improvements and strategies on planning and operation of future power systems.

Chapter 10: The rapid evolution of the passive distribution technical infrastructure towards one active was due to the emergence of new innovative technologies that improved its efficiency and reliability. These include the use of smart meters and automation devices. The increasing number of small-size renewable energy sources has represented another challenge to the policy of the distribution network operators (DNOs) to register the direction of the target regarding the minimization of greenhouse gas emissions associated with the distribution process. Numerous studies revealed that an increasing number of end-users will want to become active in the electricity sector in the next period, which can have significant implications for the system strength. The equipment manufacturers developed innovative solutions based on smart meters and other advanced devices in the active electric distribution networks (AEDNs) to provide reliable and flexible communication systems. The availability of dedicated platforms and modern technologies allowed the identification of new solutions, with a high impact on the system strength. Based on these technological evolutions, the DNOs must develop efficient strategies to improve the operating conditions of the AEDNs and reduce the environmental impact based on innovative technologies to integrate advanced techniques and real-time communication solutions.

In the chapter, the authors proposed new smart devices-based strategy in the optimal planning and operation of the AEDNs integrating the small-size local renewable generation sources with various penetration degrees to improve the system strength. Testing has been done in a Romanian AEDN, and the results confirmed obtaining the technical and economic benefits through its implementation.

The editors
Hassan Haes Alhelou, Nasser Hosseinzadeh, Behrooz Bahrani

Chapter 1

Power system strength assessment with high penetration of inverter-based resources – a conceptual approach

Nasser Hosseinzadeh[1], Lakna Liyanarachchi[1], Ameen Gargoom[1], Hassan Haes Alhelou[2], Behrooz Bahrani[2], Fateme Fahiman[3] and Ehsan Farahani[4]

Abstract

The power system strength has been conventionally evaluated by finding the short circuit ratio (SCR) at nodes of a power network. In systems with inverter-based resources (IBRs), the short circuit index is still usually used although several other indices are defined to improve its accuracy for systems with IBRs. However, even these indices do not represent – to a good accuracy – the actual value of the system strength for a system with many IBRs, which have the capability of injecting reactive power to the network in addition to absorbing reactive power from the network. There is obviously a pressing need for finding a new metric for correctly evaluating the system strength in modern power systems dominated by IBRs such as wind farms, solar power plants with storage, other renewable energy sources, microgrids, and active distribution networks. Although this can be done by detailed system modeling through EMT-based simulations, but there is a need for a metric that can evaluate system strength with a good approximation without having to do detailed modeling, similar to the way that SCR can be easily calculated. This chapter discusses the concept of power system strength in modern power grids in contrast with the system strength in conventional systems and provides the big picture of this concept. As such, it serves as an introduction to other chapters of this book as each of them will discuss some specific aspects of the power system strength notion for the power systems with IBRs.

[1]School of Engineering, Deakin University, Australia
[2]School of Electrical and Computer System Engineering, Monash University, Australia
[3]Energy Australia, Australia
[4]Australian Energy Market Operator (AEMO)

Acronyms

AEMC	Australian Energy Market Commission
AEMO	Australian Energy market Operator
CSIRO	Commonwealth Scientific and Industrial Research Organisation
EMT	electromagnetic transient
EPRI	Electric Power Research Institute
EV	electric vehicle
FACTS	flexible AC transmission system
G-PST	global power system transformation
HVDC	high-voltage direct current
IBR	inverter-based resource
PCC	point of common coupling
PoC	point of connection or point of coupling
PSS	power system stabilizer or power system stabilizer
PSstr	power system strength
RES	renewable energy system
SA	South Australia
SCR	short circuit ratio
SCR_rp	short circuit ratio with due consideration to the effect of reactive power injection
SDSCR	site-dependent short-circuit ratio
STATCOM	static synchronous compensator
SVC	static var compensator
Vic	Victoria, Australia
WSCR	weighted short circuit ratio

1.1 Introduction

As the share of renewable energy systems (RESs) in the generation of electricity is continuously increasing across the globe, new challenges arise in the operation and planning of the interconnected electrical power grid. Among these challenges are the methods of performing voltage stability analysis, and, as a related matter, the way that power system strength (PSstr) is assessed. Power system strength can be vaguely defined as the ability of the power system at any specific node to maintain its voltage during and following a sudden change of current, or equivalently a change in active or reactive power, at that node. The level of PSstr is mostly related to the sign and magnitude of sensitivity of voltage to the reactive power injections or absorptions, from generating plants or inverter-based resources (IBRs) connected to the concerned node/bus. In simple terms, it may be defined as a relative change in voltage for a change in load or generation. A power system is voltage unstable if $V-Q$ sensitivity is negative for at least one bus/feeder [1]. However, a more technical definition is needed, particularly in the context of the systems with high penetration of renewable energy systems, or more generally IBRs.

To generalize the concept presented by this chapter, the term IBRs will be used for any renewable energy system or any power supply, or more generally any active element that can at least sometimes inject active or reactive power into the system and is connected to the system through a power electronics inverter. Some examples of IBRs are wind energy systems, solar PV energy systems, hydro-power plants with induction generators, various storage systems, electric vehicles (EV) charging stations, STATCOMs, SVCs, and synchronous condensers. It is generally observed that the power system strength is reduced with high penetration of IBRs with grid following inverters. In such systems, the risk of instability, particularly voltage instability increases.

Low-power system strength makes the electrical power networks vulnerable to outages. There are reports of problems that have occurred when connecting new wind farms or solar farms to a weak point of a power system. In June 2022, it was reported that Australia's biggest wind and solar hybrid project (to date) had to be disconnected from the grid after causing repeated "flickers" on the grid in South Australia (SA). The "flickering" and "dimming" were caused by voltage fluctuations on the local grid; and, the hybrid wind and solar generator were identified as the cause [2]. It should be noted that this plant was still under commissioning when this voltage fluctuation problem occurred. Also, in SA, a major blackout happened in September 2016, as explained in the Australian Energy Market Operator (AEMO)'s report [3]. As explained by AEMO, two tornadoes damaged three high-voltage transmission lines, which were disconnected. A sequence of faults resulted in six voltage dips on the grid over a 2-min period. Eight wind farms had the protection settings of their wind turbines to withstand only a pre-set number of voltage dips (less than 6). Due to the high number of voltage dips, this protection feature was activated and resulted in a significant power reduction. A generation reduction of 456 MW occurred in a short time, in a period of less than 7 s. The sustained reduction in wind farm output caused a significant increase in the imported power flowing through the Interconnector from Victoria (Vic) to SA. Approximately 0.7 s after the reduction of output from the last wind farm, the flow on the Vic–SA interconnector reached such a level that it activated a special protection scheme that tripped the interconnector offline. This tripping cascaded to a blackout. It is believed that the low system strength in the area of wind farms was the initial cause of many voltage dips, which eventually resulted in a power blackout. After the South Australian blackout event in September 2016, the Australian Energy Market Commission (AEMC) made several rules to deliver improved system strength, inertia, and emergency frequency control [4]; and AEMO updated and revised operational actions.

Currently, PSstr is usually evaluated by finding an index, conventionally being the short circuit ratio (SCR), which is a reflection of the available short circuit current, or available fault current, at the concerned bus versus the capacity of the IBR aimed to be connected to the same bus. For a good understanding of SCR and its limitations in correctly and accurately measure the power system strength, let us go back to the roots of how this index was defined. The SCR index was originally defined by Charles L.

Fortescue in 1918 in the context of the behavior of electrical machines under fault conditions [5]. Later in 1970, in the context of high-voltage direct current (HVDC) stations connected to alternative current (AC) systems, SCR was defined as the ratio of the system short circuit capacity to the HVDC (or inverter) rated power [6]. It can be said that the value of SCR reflects a measure of how far the system operating point is from the voltage stability limit. The higher the SCR, the stronger the bus is with respect to voltage stability. In a way, it can be said that as the ability of the system that is connected to an IBR reduces for injecting current to the point of connection, the system strength goes lower at that point. In such a case, the bus voltage and its angle become more sensitive to the changes that happen on the IBR side. In contrast, a stronger system will counteract any change at the IBR side which affects the bus voltage by injecting sufficient current towards that bus; this makes the bus less sensitive to the changes in the IBR side, which eventually prevents severe voltage fluctuations at that bus. As an example, if the power supplies are connected with high impedances to the point of connection, this causes a weak system strength at that point.

However, it has been found that SCR does not represent – to a reasonable accuracy – the actual value of the PSstr for a bus that has several IBRs and other power electronic devices in its vicinity. As such, a number of other indices have been defined for the evaluation of the PSstr in the presence of IBRs, e.g., [7–12]. In [7], an index for measuring PSstr is suggested called effective SCR (ESCR), which takes into account the reactive power compensation of capacitor banks. It is used for the single-infeed HVDC systems. The shortcoming of the ESCR is that it does not take into account the mutual effects between nearby HVDC links. In [8], a hybrid multi-infeed effective short-circuit ratio (HMESCR) was introduced for evaluating hybrid multi-infeed HVDC systems. The capacity of VSC systems to provide reactive power was considered in the derivation of this index, while taking into account the mutual effects of nearby HVDC links. In [9,10], an index called weighted SCR (WSCR) is introduced, which takes into account all IBRs in an area and calculates a single index assuming that all IBRs are connected at the same PoC. Although WSCR gives more realistic values for the PSstr, but it does not include the interaction of the IBRS in the vicinity area which is not directly connected to the PoC. In [11], an index called site-dependent SCR (SDSCR) is proposed, which is based on SCR, but takes into account the interactions among the IBRs which are electrically connected to the point of connection. It is shown that SDSCR can evaluate the distance to the voltage stability limit more accurately compared to the SCR. In [12], a scheme to identify mutual-interference boundaries of IBRs for the estimation of PSstr is proposed. It aims to take into account the interactions between IBRs that are inside each boundary in determining PSstr.

To assess the power system strength correctly, dynamic simulations for the whole system should be carried out for some specific disturbances enforced on the power system. Usually, for a good accuracy, software packages based on electromagnetic transient (EMT) simulations such as PSCAD are used in practice. This method for system strength assessment requires a high level of computational requirement and takes a relatively long time.

Online identification of power system strength is required to properly operate the system with high penetration of IBRs to keep the voltage quality across the system and to prevent instability and outages. In Australia, the Commonwealth Scientific and Industrial Research Organisation (CSIRO) has put online identification of system strength among high-priority research topics in a report on research for the global power system transformation (G-PST), which was done through Electric Power Research Institute (EPRI) [13]. Identification of the system strength in real-time operation in a way that can take the IBR dynamics into the evaluation method is indicated as one of the demanded areas of research. In a report by Australian Electricity Market Operator (AEMO) [14], one of the requirements for a system with IBRs is indicated as the determination of the criteria for a stable voltage waveform. This requirement should be further reflected in the efficient level of the system strength, which is meant to be a level of power system strength that makes a stable voltage waveform at the point of common coupling (PCC) of IBRs. When a new renewable energy system (RES), such as solar PV or wind energy system, requests connection to the power grid, it is necessary to perform studies to ensure that its addition to the system will not weaken the system to a point that voltage fluctuations are aggravated when a disturbance occurs in the vicinity of the PCC of that RES.

As the determination of the power system strength with the inclusion of IBR dynamics requires EMT simulations, which take long time due to excessive computations, there is a demand for fast schemes for power system strength assessment. The fast screening is proposed as the first attempt for screening buses of interest in the system to determine the critical buses, which would need further scrutiny. Once critical buses are identified through the fast system strength assessment technique, then full EMT-based simulations will be carried out only for the critical buses.

Accordingly, this research proposes ways of finding novel schemes for fast assessment of power system strength with the inclusion of IBRs. The proposed schemes will be developed at the system level to provide a holistic identification of critical buses with regard to their level of power system strength.

1.2 Work in progress for determination of power system strength in a large grid

This study is carried out in two directions: (1) reduction of the system dynamic model by partitioning it into two parts: the section of the system that contains the buses where the new RESs will be connected (hereby called the system section under study), and the rest of the system. Then, an equivalent dynamic model of the rest of the system will be worked out; and only the section under study will be modeled with details. (2) A new method of assessing the power system strength will be worked out that takes into account the inclusion of IBRs, their interactions, and also other system components such as synchronous condensers, and FACTS devices such as SVCs and STATCOMs.

In this book, Chapter 4 addresses direction (1) and Chapter 2 addresses direction (2).

The methodology of this study is explained in the following sections. In Section 1.2.1, the conventional assessment of PSstr and its extension to the systems with IBRs is reviewed conceptually, and an alternative index named as SCR_rp (which refers to *SCR* taking the effect of *reactive power* injected or absorbed) is introduced to determine PSstr more accurately than SCR when IBRs are set in their var control mode. In Section 1.2.2, a relationship between SCR and power system voltage stability is discussed. Section 1.2.4 presents an outline of a new method for assessing power system strength. Finally, Section 1.2.5 provides the summary and future directions.

1.2.1   Method of assessing power system strength

The term short circuit ratio, SCR, was first introduced by Charles L. Fortescue, who defined SCR as the ratio of the reactance of a synchronous machine to its own synchronous impedance. This definition can be found in his seminal paper "Method of Symmetrical Co-Ordinates Applied to the Solution of Polyphase Networks," which was published in the Transactions of the American Institute of Electrical Engineers in 1918. In Section III of the paper, Fortescue writes: "The short-circuit ratio for any machine is defined as the ratio of the reactance of the machine to its own synchronous impedance. This ratio is important in the study of symmetrical polyphase systems because it is a measure of the strength of the machine under short-circuit conditions."

In fact, SCR of a synchronous generator is defined as the ratio of the field current required to generate rated voltage on an open circuit to the field current required to circulate rated armature current on a short circuit. A high value of SCR for a synchronous machine would mean a better voltage regulation and a higher steady-state stability limit for the machine, but higher short circuit current in its armature winding. SCR for a synchronous machine can be derived as given in (1.1):

$$\text{SCR for a synchronous machine} = \frac{I_f \text{ for rated O.C. voltage}}{I_f \text{ for rated S.C. current}} \cong \frac{1}{X_d \text{ in pu}} \quad (1.1)$$

where X_{d} is the per unit value of direct axis synchronous reactance.

A synchronous machine with a low value of SCR is sensitive to the load variations, i.e. a large variation in terminal voltage occurs due to a change in load. To prevent variation of the terminal voltage, the field current of the machine is to be varied over a wide range. Also, for a small value of SCR, the synchronizing power of the machine is small. The synchronizing power keeps the machine in synchronism, thus, the synchronous machine with a low value of SCR has a low stability limit. Therefore, a synchronous machine with a low SCR is less stable when operating in parallel with other machines. A synchronous machine with a high value of SCR has a better voltage regulation and higher steady-state stability limit, but with a higher short circuit fault current in its armature, with a bigger size and higher cost.

Hydro and salient pole machines have smaller direct-axis inductance, i.e. smaller X_{d}, which is the result of larger air gaps in generator design. This results in higher values of SCR. Such a machine will be loosely coupled to the grid, and its response will be slow. This increases the machine's stability while operating on the grid. On the other hand, the smaller the SCR, the larger would be the machine's inductance. Such a machine, e.g. a turbo generator or cylindrical rotor machine, will be tightly coupled to the grid, and its response will be fast. This reduces the machine's stability while operating on the grid.

SCR was also defined from the power system perspective. It was first defined in the context of HVDC links connected to AC systems. In 1970, [6] defined the SCR of an AC power system at the connection point of a HVDC station as the ratio of the AC system short-circuit power to the HVDC station-rated DC power. This definition with some variations has been used to date for various applications. It should be noted that this SCR is related to the strength of the system side in contrast to the strength of the machine, which was given by (1.1). The SCR of an AC system connected to an HVDC link was defined as:

SCR for AC system connected to HVDC

$$= \frac{\text{Short circuit MVA of AC system at the PoC}}{\text{DC converter MW rating}} \tag{1.2}$$

It should be noted that this definition is based on fundamental frequency quantities. The article [6] has further investigated under which SCR values large harmonic voltages occur in the AC bus voltage of an HVDC station, which is called harmonic instability. With some benchmark testing, the article concluded that harmonic instability can be avoided by keeping SCR > 2 by appropriate settings of the control circuits. Later, the IEEE standard 1204-1997 [7] suggested the following categories for SCR values in AC/DC systems, which seems to be based on [15], which is itself based on some CIGRE and IEEE studies in the 1980s [16–18].

(a)   A **high SCR** AC/DC system is categorized by an SCR value greater than 3.
(b)   A **low SCR** AC/DC system is categorized by an SCR value between 2 and 3.
(c)   A **very low SCR** AC/DC system is categorized by an SCR value lower than 2.

As a summary of the previous studies, these categories for the level of system strength were suggested by [15] at a conference in 1991. These classifications of system strength are based on the relationship between the maximum available power (MAP) of the DC/AC inverter and the required power for the specified system conditions. The paper states, "Based on an average inverter data and on the assumption that the inverter would be compensated to operate at the unity power factor," the following approximate values of SCR are applicable:

Strong – High SCR system: SCR > 3
Weak – Low SCR system: 3 > SCR > 2
Very weak – Very low SCR system: SCR < 2
Critical SCR: Approximately equal to 2 in

It should be emphasized that *this classification is based on the inverter operating at the unity power factor*.

Further to the definition of SCR, highlighted some important points in relation to the use of SCR in analyzing the system strength in the context of the interactions between AC and DC systems of HVDC links. A system consisting of many generators, transmission lines, and loads, experiences different values of SCRs as

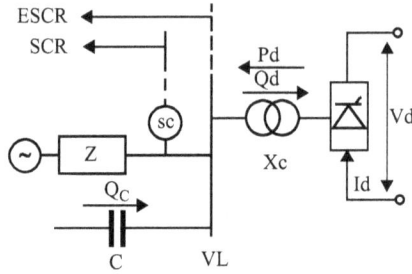

Figure 1.1   Simplified representation of a DC link feeding an AC system with shunt capacitors (courtesy of [15])

changes occur in the system configuration by switching generators, lines, transformers, etc., and loads change almost continuously. As such, this paper defined other indices related to the SCR, namely effective SCR (ESCR), operating SCR (OSCR), and operating effective SCR (OESCR). The model which was used by [15] for the system strength analysis is shown in Figure 1.1.

As renewable energy systems started penetrating into the power grids, the definition of SCR in (1.2) was extended to a power system connected to any IBR system as given in (1.3):

SCR of AC system connected to any IBR

$$= \frac{\text{Short circuit MVA of AC system at the PoC}}{\text{Power rating of IBR}} \tag{1.3}$$

Although there are some differences between AC/DC systems in HVDC links and AC/IBR systems, more or less, the same categories of SCR values given by the IEEE standard 1204-1997 have been used to determine the level of system strength at the PoCs of AC/IBR systems, i.e., **SCR < 2** representing a very weak system, **2 < SCR < 3** for a weak system, and **SCR > 3** representing a high level of system strength. However, it has been observed that there are cases of power systems connected to IBRs where SCR shows a value equal or even bigger than 3, but sustained voltage fluctuations occur after a fault inception. Of particular interest are the effect of control loops of IBRs and the effect of interactions among IBRs installed in the vicinity of PoC on the voltage variations after the inception of any major disturbance close to the PoC. It is obvious that SCR as defined in (1.3) cannot capture the true level of the system strength. As such, a new approach to assessing the level of PSstr is required. Although detailed EMT simulations can find dynamic responses of the system at any node, and this can be utilized to find the weak points in the system, but carrying out EMT simulations for all nodes of a large power system, also called bulk power system (BPS), is a forbidden practice! Therefore, we are working on a new approach to find weak points in the system through a fast screening of the level of system strength before we apply EMT simulation studies for those identified weak nodes for a more thorough investigation.

Some observations:

- This definition of SCR as an index for assessing the power system strength is meaningful for a converter-based system that needs to absorb reactive power from the system for its operation.
- Therefore, for the modern converters which may be controlled to provide reactive power to the system, SCR may not be considered as a proper index for assessing the system's strength. For example, increasing the converter-rated power results in a lower value of SCR, which means the power system becomes weaker compared to the system strength when the converter-rated power was lower. But, this would only be true if the IBR was a receiver of reactive power. For the modern smart converters, this is not true.
- SCR is a static index. It cannot capture the dynamics of PLL, control loops, and IBR interactions.
- For fast screening, we may still use a static analysis approach for developing a power system strength index which would reflect the behavior of the system point of connection to IBRs more accurately in terms of its stiffness related to voltage fluctuations after the inception of a disturbance.

Many attempts have been made to address the mentioned issues and several indices are proposed such as ESCR, OSCR, OESCR [15], WSCR proposed by the Electric Reliability Council of Texas (ERCOT) [9,10], composite SCR (CSCR) proposed by General Electric (GE) [19], SCR with interaction factors (SCRIF) [20], site-dependent SCR (SDCSR) [11], and inverter interaction level SCR (IILSCR) [21]. We define a modified SCR, which takes into account the effect of reactive power injected or absorbed by the IBR and any other flexible devices that can inject reactive power into the PoC, such as SVCs and other FACTS devices. This proposed index is hereby called short circuit ratio with the effect of reactive power, which is abbreviated as SCR_rp, as follows:

$$\text{SCR}_{\text{rp}} = \frac{S_{\text{sc}} \pm Q_{\text{dev}}}{P_{\text{IBR}} \mp Q_{\text{IBR}}} \tag{1.4}$$

where S_{sc} is the **minimum** short circuit capacity of the power system connected to the PoC when the IBR is not connected, P_{IBR} is the **operational** active power of the IBR connected to the PoC, Q_{dev} is the reactive power injected by or absorbed by any flexible device connected to the PoC, and Q_{IBR} is the reactive power injected by or absorbed by the IBR, **all in per unit**. For Q_{dev}, the + sign is used when the device is injecting reactive power, and, for Q_{IBR} the − sign is used when the IBR is injecting reactive power (as it appears in the denominator of the SCR).

The SCR_rp, as defined here, is a **static index** for determining the PSstr with due consideration to the effect of reactive power control of IBRs and also the effect of any other source of reactive power, which either injects or absorbs reactive power to/from the PoC. There is a similarity between the ESCR and SCR_rp, but they are not the same. The index ESCR was introduced by [15] in 1991 in the context of AC/DC (HVDC) systems, where SCR and ESCR were defined as follows:

$$\text{SCR} = \frac{S_{sc}}{P_{dc}} \tag{1.5}$$

$$\text{ESCR} = \frac{S_{sc} - Q_C}{P_{dc}} \tag{1.6}$$

where S_{sc} is the minimum value of three-phase short circuit MVA of the ac system corresponding to the transmitted power; P_{dc} is the dc power; Q_C is the Mvar of all shunt capacitors, including AC filter capacitors, connected at the convertor station AC bus, expressed in per unit of P_{dc}.

Equation (1.5) basically defines SCR as the ac system admittance expressed in per unit of dc power. Then, [15] puts an argument about the destabilizing effect of the filter shunt capacitors:

"It is easy to appreciate the destabilizing effect of capacitors by considering an ac voltage reduction for any reason: capacitors will generate less vars at the lower voltage at the very moment when more vars are required to counteract the original voltage reduction."

This argument is about the effect of shunt capacitors in reducing the system strength when the voltage reduces for any reason. Based on this, the article defines the effective short circuit ratio ESCR in such a way that it gives a smaller value than SCR. In comparing our proposed SCR_rt with ESCR, it is important to understand that ESCR gives the level of system strength to reflect the higher vulnerability of the system to voltage instability due to the fixed **passive** shunt capacitors at the AC side of the DC/AC inverter. The effect of these fixed passive capacitors is taken as reducing SCR.

However, we are defining SCR_rp when **active devices** are injecting (or sometimes absorbing) reactive power. These active devices, including the IBR itself, do not behave similar to passive shunt capacitors; their injected reactive power does not necessarily decrease as the voltage decreases. Therefore, not only these active devices do not have a destabilizing effect, but they actually increase the level of system strength compared to a situation when they do not exist, or they do not provide any reactive power. This is the reason that we use a positive sign for Q_{dev} when it is injecting reactive power in contrast with a negative sign used in the ESCR. In fact, SCR_rp generates a higher value compared to SCR while ESCR generates a lower value.

Further to the definition of ESCR, [15] defines an operational SCR based on the following consideration:

When considering operating conditions at other load values, usually lower than rated load, the corresponding operational SCR changes compared to the one found for the rated load. In finding the operating SCR or operating ESCR, which are abbreviated as OSCR or OESCR, actual values of shunt capacitors and power level, not the rated power, must be used in the formulas (1.5) and (1.6).

Therefore, it is very important to understand that as the configuration or operating conditions of a BPS change, the power system strength at any bus of the system changes. As such, in assessing the power system strength, we should not use a fixed value of SCR, rather operational SCR needs to be used. Therefore, all quantities used in (1.4) to calculate SCR_rp shall be operation values in per unit.

Before we reach to more complicated matters on the assessment of the power system strength in dynamic situations, let us illustrate the difference between SCR_rp and SCR, by using a simple system comprised of an IBR connected to an infinite bus via a transformer and a transmission line, as shown in Figure 1.2.

For the system shown in Figure 1.2, we have used an actual solar farm, but an infinite bus is used to represent the rest of the power grid. The IBR is set in the reactive power control mode. For three scenarios, i.e. IBR at unity power factor ($Q = 0$), IBR injecting 0.2 pu of reactive power, and IBR absorbing 0.2 pu reactive power, we have calculated SCR and SCR_rp. For the same scenarios, we have found the voltage responses at the PoC when a two-phase to-ground fault happens at the end of the transmission line through a PSCAD simulation of the system. The voltage responses for the three scenarios are shown in Figure 1.3.

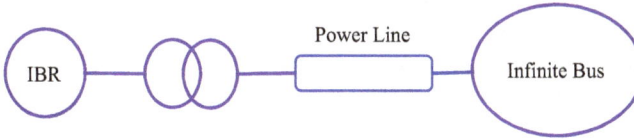

Figure 1.2   *An IBR connected to an infinite bus via a transformer and a transmission line*

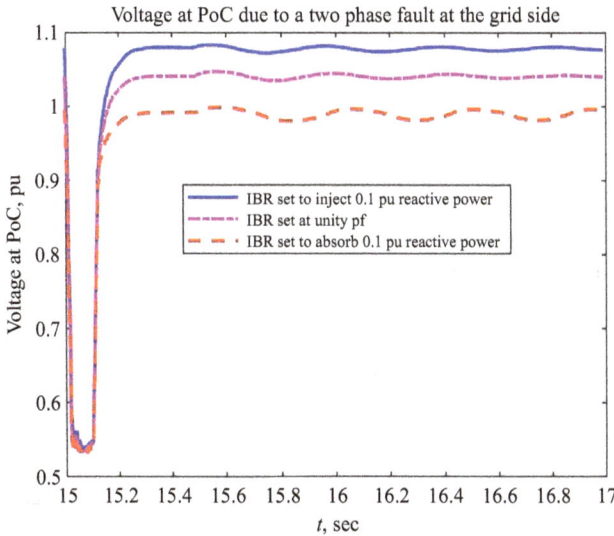

Figure 1.3   *(1) Red dashed plot, IBR absorbs Q; (2) pink _. plot, Q = 0; (3) blue solid plot, IBR injects Q*

As seen in Figure 1.3, when the IBR absorbs reactive power, the voltage at the PoC has an oscillatory response. It is clearly observed that voltage fluctuations are different in the three scenarios; sustained fluctuations are less in the cases when the IBR operates at a unity power factor or injects reactive power into the grid. It is observed that the voltage response to a major fault is more stable when the IBR injects reactive power into the grid compared to when it is absorbing reactive power. Also, the unity power factor operating makes the voltage more stable than when the IBR absorbs reactive power. Therefore, it is noticed that the PoC will have more power system strength when the IBR injects reactive power compared to other cases.

The index SCR cannot distinguish the differences between these three cases: for the three cases, SCR will be the same!

On the other hand, our proposed SCR_rp takes into account the differences caused by the direction and value of reactive power from the IBR. The SCR_rp will be higher when the IBR injects reactive power to the grid, and lower when it absorbs reactive power.

Furthermore, if there is any other flexible device connected to the PoC which can inject or absorb reactive power, its effect will be taken into account in the value of SCR_rp.

Here are the SCR and SCR_rp values found for the three cases, assuming that both the transformer and the transmission lines have negligible resistances, and their combined series reactance is X_{Po-CIB} (to be read as the reactance between PoC and the infinite bus), which in this example is equal to 0.35 pu:

SCC at PoC $= 1/X_{PoC-IB} = 1/0.4 = 2.5$ pu

SCR $=$ SCC$/S_{IBR} = 2.5/(80/100) = 3.125$

With this level of SCR, we expect that the system is strong, and there should be no voltage fluctuations after the clearance of the fault. It is seen from Figure 1.3 that when $Q = 0$ or $Q = 0.2$ pu, the voltage response shows a relatively strong system although there is a short-term voltage fluctuation just after the fault clearance. In fact, this short-term voltage fluctuation is smaller in the case of $Q = 0.2$ compared to $Q = 0$. However, when $Q = -0.2$, there is a visible voltage fluctuation long after the fault clearance. The important point is that SCR cannot see any difference among these three cases. In all cases, it is telling us that the system strength is high, but this is obviously not correct as we are observing sustained oscillations in the case when the IBR is absorbing reactive power.

Now, let us find the value of the proposed SCR_rp for the three cases. It should also be noted that in this example, we are assuming that there is no shunt device at the PoC that would generate reactive power when the voltage goes down, i.e., we have assumed $Q_{dev} = 0$. Using (1.4),

$$\text{SCR}_{rp} = \frac{S_{sc} \pm Q_{dev}}{P_{IBR} \mp Q_{IBR}}$$

When $Q_{IBR} = 0$, SCR_rp = SCR = 3.125.

When $Q_{IBR} = 0.2$ pu,

$$\text{SCR}_{\_rp} = \frac{2.5}{0.8 - 0.2} = 4.2$$

In this case, SCR_rp gives us the indication that the system is very strong; and this is validated by the quick damping of voltage oscillations in this case, as shown in Figure 1.3.

When $Q_{IBR} = -0.2$ pu,

$$\text{SCR}_{rp} = \frac{2.5}{0.8 + 0.2} = 2.5$$

In this case, SCR_rp says that the system strength is low. Sustained voltage fluctuations in this case, shown in Figure 1.3, validate this value, as well. Of course, we do not expect that any static index will be able to predict the dynamic response of the system exactly. The main point here is that our proposed index can successfully follow the change in responses as the level of reactive power injected or absorbed by the IBR changes. This makes it superior to SCR, which is still commonly used for the assessment of power system strength, at least in the screening phase.

1.2.2 Relationship between SCR and power system voltage stability

To illustrate the concept of static voltage stability and the relationship of SCR to this concept, usually a simple two-bus system is analyzed. Let us consider a power system, approximated as a source connected to an infinite bus, connected to bus 1, supplying power to a load at bus 2 via a transmission line with an impedance $\mathbf{Z}_{12} = \mathbf{Z}_{line} = R_{line} + jX_{line}$. The load has an impedance of $\mathbf{Z_L} = R_L + jX_L$ and consumes a complex power of $\mathbf{S_L} = P_L + jQ_L$. This system is shown in Figure 1.4. Bus 1 is the swing bus, and so, V_1 and δ_1 are fixed. With a simple circuit analysis, we find the following. Also, the transmission line has a fixed impedance with a given X/R ratio. For the load, usually a constant power load is used; here, for the

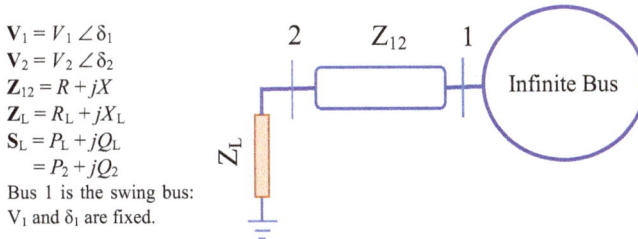

$\mathbf{V}_1 = V_1 \angle \delta_1$
$\mathbf{V}_2 = V_2 \angle \delta_2$
$\mathbf{Z}_{12} = R + jX$
$\mathbf{Z_L} = R_L + jX_L$
$\mathbf{S_L} = P_L + jQ_L$
$\quad\ = P_2 + jQ_2$
Bus 1 is the swing bus:
V_1 and δ_1 are fixed.

Figure 1.4   A simple two-bus system to illustrate the voltage stability concept

sake of illustration, we have given the load impedance, but it can vary to change P_L and Q_L as desired. Bold letters are used for complex quantities, and regular fonts for magnitudes. Also, all quantities will be in per unit on a common system base. With these in mind, let us do a simple analysis as follows:

$$Z_{12} = R_{12} + jX_{12}$$

$\frac{X}{R}$ ratio of transmission line $= \frac{X_{12}}{R_{12}} \triangleq a$

Power factor of the load $= \cos(\theta_{Z_L}) = \frac{P_L}{S_L} = PF$

$$\tan(\theta_{Z_L}) = \frac{Q_L}{P_L} = \tan\left(\cos^{-1}(PF)\right) = \frac{X_L}{R_L} \triangleq b$$

$$I_{12} = \frac{V_1}{Z_{12} + Z_L} = \frac{V_1}{(R_{12} + jX_{12}) + (R_L + jX_L)}$$

$$I_{12} = \frac{V_1}{(R_{12} + ja(R_{12})) + (R_L + jb(R_L))} = \frac{V_1}{(R_{12} + R_L) + j(a(R_{12}) + b(R_L))}$$

It is noticed that the current depends on a few factors: the voltage at the system terminal, the line impedance and its X/R ratio, the load, and its power factor.

The voltage at the load terminal will be

$$V_2 = V_1 - Z_{12}I_{12} = V_1 - \frac{Z_{12}V_1}{(R_{12} + R_L) + j(a(R_{12}) + b(R_L))}$$

$$V_2 = V_1\left(1 - \frac{R_{12} + ja(R_{12})}{(R_{12} + R_L) + j(a(R_{12}) + b(R_L))}\right) \tag{1.7}$$

Also, from the load side, we have:

$$S_L = P_2 + jQ_2 = V_2(I_{12}^*) = \frac{V_2}{Z_L} = \frac{V_2}{R_L + jb(R_L)} \tag{1.8}$$

If we have the line impedance, by combining (1.7) and (1.8), we can find V_2 as a function of P_2 (or S_L) for different values of load power factor. Then a plot of V_2 versus P_2 is called the P–V curve. A set of PV curves for lagging, unity, and leading load power factor values are shown in Figure 1.5. The voltage at bus 1, $V_1 = V_1\angle\delta_1$, is assumed to have constant magnitude and angle.

The following points are revealed by the P–V curves.

- For load-lagging PF or unity PF, V_2 decreases as P_2 increases.
- For load-leading PF, it is possible that V_2 increases as P_2 is increased from zero value (open circuit at load side), but then decreases as P_2 increases towards its maximum possible value, $P_{2,max}$.
- The value of $P_{2,max}$ depends on the PF of the load. Usually, it is higher for a capacitive (leading PF) load than for an inductive (lagging PF) load.
- Normally, the system operates at a much less P_2 than the maximum possible value of P_2, i.e. P_2 is much less than $P_{2,max}$ ($P_2 \ll P_{2,max}$).

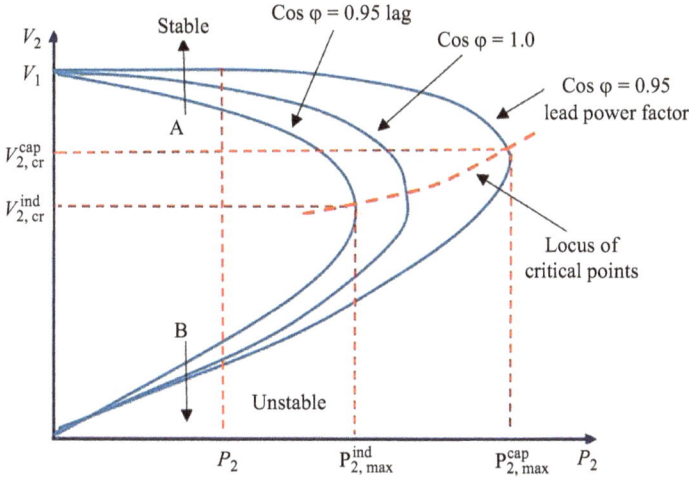

Figure 1.5   Load P–V curves with various values of power factor and for a constant V_1 at the system side

- The system can stay stable only up to $P_{2,\text{max}}$, any attempt to further increase the power consumed by the load will result in voltage collapse. The point on the curve corresponding to $P_{2,\text{max}}$ is called the *critical point* or *voltage collapse point*. At this point, the voltage will have the value of $V_{2,\text{cr}}$.

Now, let us find SCR, and find out the relationship between SCR and *P–V* curves:

$$I_{sc} = \frac{V_1}{Z_{12}} = \frac{V_1}{(R_{12} + jX_{12})} = \frac{V_1}{(R_{12} + ja(R_{12}))}$$

Short circuit capacity at bus $2 = \text{SCC} = I_{sc} V_2$

There are different ways of defining the SCR:

$$SCR_{\text{rated}-S} = \frac{SCC}{S_L} \text{ where } S_L \text{ is the load} - \text{rated apparent power}$$

$$SCR_{\text{rated}-P} = \frac{SCC}{P_L} \text{ where } P_L \text{ is the load} - \text{rated active power}$$

$$SCR_{\text{operating}-S} = \frac{SCC}{S_{L-op}} \text{ where } S_{L-op} \text{ is the load} - \text{operating apparent power}$$

$$SCR_{\text{operating}-P} = \frac{SCC}{P_{L-op}} \text{ where } P_L \text{ is the load} - \text{operating active power}$$

Each one of these has their merits. However, it is clear that the system strength is not a fixed value for any node of the system; it changes as the system configuration and operation changes. Furthermore, whether the apparent power of the

IBR should be used or its active power, we used the latter as in the definition of our proposed SCR, we deal with the reactive power of the IBR separately. As such, here we derive $\text{SCR}_{\text{operating}-P}$

$$\text{SCR}_{\text{operating}-P} = \frac{SCC}{P_{L-op}} = \frac{I_{sc}V_{2-op}}{P_{L-op}} \tag{1.9}$$

$$I_{op} = \frac{V_{2-op}}{Z_{L-op}} = \frac{V_{2-op}}{R_L + jb(R_L)}$$

$$S_{L-oP} = V_2 I_{op}^* = \frac{V_2 V_2^*}{R_L - jb(R_L)} = \frac{V_2^2}{R_L - jb(R_L)} \tag{1.10}$$

By using (1.9) and (1.10), we can calculate the SCR for each operating point.

Example 1: In Figure 1.4, if we have $\mathbf{V}_1 = 1\angle 0°$ pu, $\mathbf{Z}_{12} = 0.06 + j0.6$ pu, and for a specific operating point $\mathbf{Z}_L = 1.5 + j0.3$ pu, find the SCR.

Solution:

$$a = \frac{X_{12}}{R_{12}} = \frac{0.6}{0.06} = 10$$

$$b = \frac{X_L}{R_L} = \frac{0.3}{1.5} = 0.2$$

$$V_2 = V_1 \left(1 - \frac{R_{12} + ja(R_{12})}{(R_{12} + R_L) + j(a(R_{12}) + b(R_L))} \right)$$

$$V_2 = 1\angle 0 \left(1 - \frac{0.06 + j(10)(0.06))}{(0.06 + 1.5) + j[(10)(0.06) + 0.2(1.5)]} \right)$$

$$V_2 = 1 - \frac{0.06 + j0.6}{1.56 + j0.9} = 0.8047 - j02719 \text{ pu}$$

$$S_{L-oP} = \frac{V_2^2}{R_L - jb(R_L)}$$

$$S_{L-oP} = \frac{V_2^2}{1.5 - j0.3} = \frac{(0.8494)^2}{1.5 - j0.3} = 0.4624 + j0.0925 \text{ pu}$$

$$P_{L-op} = real(S_{L-oP}) = 0.4624 \text{ pu}$$

$$I_{sc} = \frac{V_1}{Z_{12}} = \frac{1\angle 0}{0.06 + j0.6} = 0.1650 - j1.6502 \text{ pu}$$

$$SCC = I_{sc}V_{2op} = (1.6584)(0.8494) = 1.4086 \text{ pu}$$

$$SCR_{\text{operating}-P} = \frac{SCC}{P_{L-op}} = \frac{1.4086}{0.4624} = 3.0 \text{ pu}$$

At this specific operating point, the bus at the load terminal has a relatively high strength. As illustrated in this example, SCR can have different values in different operating conditions even for the same node. Now, let us investigate a few points about the usual definition of SCR and critical SCR (SCR$_{cr}$) [11,22].

Most commonly, SCR at bus i of a power system, where an IBR is connected, is defined as:

$$\text{SCR}_i = \frac{SCC_i}{P_{IBR,inj}} \tag{1.11}$$

where SCC_i is the short circuit capacity of the system at bus i (when IBR is not connected to bus i and the bus is short circuit), and $P_{IBR,inj}$ is the capacity of the IBR for injecting power to bus i. Note that both SCC_i and $P_{IBR,inj}$ are defined based on the rated values, not operational values. This is an indication of the level of the system strength when the IBR operates at its full capacity. If we use the minimum level of the SCC_i and rated value of $P_{IBR,inj}$ to find SCR_i, then it gives a conservative value of the system strength because, in this case, usually during the system operation SCC_i will be higher and $P_{IBR,inj}$ will be lower, so SCR_i will have a higher value than the conservative value. The short circuit capacity in (1.11) can be found as follows:

$$\text{SCC}_i = \left| V_i I_{i,sc}^* \right| = \frac{V_i^2}{Z_{Th,ij}} \tag{1.12}$$

where V_i is the rated voltage at bus i and $Z_{Th,ij}$ is the equivalent Thevenin impedance between bus i and bus j, at a node where the system is strong and can be approximated by an infinite bus. By combining (1.11) and (1.12), the "rated" value of SCR can be written as:

$$SCR_i = \frac{V_i^2}{\left(Z_{Th,ij}\right)\left(P_{IBR,inj}\right)} \tag{1.13}$$

Based on (1.11), if the IBR could inject a power equal to the short circuit capacity of the system at bus i, i.e. $P_{IBR,inj} = SCC_i$, then SCR_i would be equal to 1. However, usually, $P_{IBR,inj} \ll SCC_i$, and as such, the expectation is that $SCR_i \gg 1$. The bigger SCR_i, the stronger the system. Therefore, we can consider the value of SCR_i as a measure of the distance of bus i to the voltage collapse point.

There are some confusions in the literature with regard to the definition of the critical SCR and its value. In some references, critical SCR is defined as the SCR value at the voltage collapse point. We would suggest that this level of SCR be called the short circuit ratio at the voltage collapse point ($SCR_{V\text{-}collapse}$) rather than critical SCR because we believe SCR will be a critical value when the system operates at the lowest allowable voltage value. We suggest the SCR at the lowest allowable voltage value be denote as SCR_{cr}.

$SCR_{cr} = SCR_i$ from 1.13 when the voltage at bus i is at its minimum allowable value.

The IEEE standard 1204-1997 [7] defines the critical SCR as the value of SCR that indicates the system being on the verge of voltage instability. In other words, it would be the value of SCR at the voltage collapse point. Based on [22], this value of SCR is given by (1.14):

$$SCR_{V-collapse,i} = \frac{2}{S_{IBR}} \left[-\frac{P_{IBR} + aQ_{IBR}}{\sqrt{1+a^2}} + S_{IBR} \right] \qquad (1.14)$$

where a is the X/R ratio of the impedance between the system bus and the IBR bus, i.e. $Z_{TH,12}$ in Figure 1.4. As seen from (1.14), the value of SCR at the voltage collapse point, i.e. static voltage stability limit, depends on the apparent power generated by the IBR, its power factor (which determines P_{IBR} and Q_{IBR} once S_{IBR} is known, and X/R ratio of $Z_{TH,12}$). Here are some observations deduced from (1.14). It should be noted that the operational SCR, i.e. SCR_{op}, should always be bigger than $SCR_{V-collapse}$ for stable operation of the system. Therefore, in all the following scenarios,

$$SCR_{op} > SCR_{V-collapse,i} \text{ for stable operation}$$

- When $Z_{TH,12}$ is purely resistive, i.e. $a = 0$,

$$SCR_{V-collapse,i} = \frac{2}{S_{IBR}} \left[-\frac{P_{IBR} + aQ_{IBR}}{\sqrt{1+a^2}} + S_{IBR} \right] = 2(1 - P_{IBR}/S_{IBR})$$

In this case, depending on the power factor of the IBR, which determines $\frac{P_{IBR}}{S_{IBR}}$, knowing that $0 \le P_{IBR} \le S_{IBR}$, $SCR_{V-collapse}$ will be in the range of 0–2.

- When $Z_{TH,12}$ is purely inductive, i.e. a tends to infinity,

$$SCR_{V-collapse,i} = \frac{2}{S_{IBR}} \left[-\frac{aQ_{IBR}}{a} + S_{IBR} \right] = 2(1 - Q_{IBR}/S_{IBR})$$

In this case, depending on the power factor of the IBR, we have three possibilities:

- IBR operates at unity power factor, then $SCR_{V-collapse,i} = 2$. This means that the operational SCR must be bigger than 2.
- IBR injecting reactive power, then $0 \le Q_{IBR} \le S_{IBR}$, $SCR_{V-collapse}$ will be in the range of 0–2. This means that the operational SCR must be bigger than 2.
- IBR absorbing reactive power, then $-S_{IBR} \le Q_{IBR} \le 0$, $SCR_{V-collapse}$ will be in the range of 2–4. The worst case would be when the IBR only absorbs reactive power and produces no active power, then the value of $SCR_{V-collapse}$ approaches 4. This means that the operational SCR must be bigger than 4 for this worst case to maintain system stability.

From our perspective, the best way of defining the critical SCR, denoted as SCR_{cr}, is paying attention to what would be considered as a critical status of the system in operation. We would suggest that this be defined based on the allowable lowest stable voltage at bus i given the load power factor.

Example 2: If we know that an IBR is operating at unity power factor and is generating 0.7 pu of active power. The impedance between IBR and the infinite bus is $0.05 + j0.45$ pu. The lowest acceptable voltage at bus i would be 0.9 pu. What would be the critical SCR in this case?

Solution:
Using (1.13),

$$SCR_i = \frac{V_i^2}{\left(Z_{Th,ij}\right)\left(P_{IBR,inj}\right)}$$

$$SCR_{i,cr} = \frac{0.9^2}{\left(|0.05 + j0.45|\right)(0.7)} = 2.56$$

So, in this case, any SCR less than 2.56 will make the voltage to go lower than 0.9 pu. This means that the maximum allowable power injection by the IBR in this case is 0.7 pu.

1.2.3 Effect of IBR dynamics on power system strength assessment

Neither of the static indices defined to date (June 2023), including SCR_rp defined in this chapter, can capture the behavior of IBR control loops and IBR interactions in a real large interconnected system with high penetration of IBRs. Ongoing research is required, and we are currently contributing to it, to solve this problem. To develop a new method for the assessment of PSstr in a system with IBRs, we need to return to the basic scientific concepts. A promising method is to carry out sensitivity analysis. It is known that the level of system strength is determined by the level of sensitivity of voltage magnitude and angle to changes in active and reactive power flow from or to the PoC under study. The level of PSstr required for the IBRs to operate in a stable manner depends on the type and technology of the inverter, as well. Grid-following inverters need higher system strength to operate in a stable manner, compared to synchronous machines or grid-forming inverters. Current-controlled inverters used in grid-following inverters rely on a phased-locked loop (PLL) to stay synchronized to the fundamental component of the voltage waveform. The PLL controls the magnitude and phase of output current to inject desired active and reactive power into the network. During faults, the suppressed voltage magnitude and distorted waveform can cause these inverters to lose synchronism with the power grid, which can result in the disconnection of IBRs. There are three main factors that influence the level of system strength in IBR-rich power networks, as follows [23]:

- The concentration of multiple IBRs in close electrical proximity to each other.
- The lack of sufficient synchronous machines support.
- The electrical distance of IBRs from major generation and load centers.

Due to control interactions in IBRs, new types of stability problems that appear over a wide range of frequencies are becoming more common. In [24], a data-driven tool called the impedance approach is used for analyzing the impacts of IBRs on the stability of power systems. This Chapter has explained various methods for obtaining the impedance responses of IBRs, including direct measurement, dynamic-event data-based approach, and estimation from black-box EMT models. Research on the impedance methods to make them suitable for grid-level stability studies is ongoing to address key hurdles by combining the impedance methods with modal analysis approaches and developing automation tools. Similarly, the development and the use of appropriate modeling and monitoring tools are fundamental for the assessment of various system strength challenges. North American Electric Reliability Corporation (NERC) recommends [25] that coordination and validation of short-circuit planning cases and the production of comparable fault currents at boundary buses between neighboring areas are needed for the continued reliability of the bulk power systems. If a large difference exists in short-circuit values between neighboring entities boundary buses, then there is a potential risk of relay maloperation and also in assessing the level of PSstr. Therefore, it is necessary to correct the inaccuracies in power system models.

1.2.4   Outline of a new method for assessing power system strength

When a request is lodged from a new RES for connection to the power grid, it is necessary to perform studies to ensure that its addition to the system will not weaken the system to a point that voltage fluctuations are aggravated when a major disturbance occurs in the vicinity of the point of connection (PoC) of that RES. To assess the power system strength correctly, dynamic simulations for the whole system may be carried out for some specific disturbances enforced on the power system. Usually, for a good accuracy, software packages based on electromagnetic transient (EMT) simulations such as PSCAD are used in practice. This method for system strength assessment requires a high level of computing capacity and takes a relatively long time. Therefore, there is a demand for fast schemes for power system strength assessment to screen all buses of the system to identify critical buses which are classified as weak and are vulnerable to voltage fluctuations/instability before a full-scale EMT simulation is performed. The detailed EMT simulations will be applied to study those critical buses, rather than for all buses of the system, in more detail and with more accuracy.

This is the outline of our approach, which is ongoing and in progress.

Usually, the focus of the PSstr study is one area of a large power system; this area is called "study area." All other areas of the large power system will be called "external area." The following steps are carried out to determine the critical buses of the study area in terms of PSstr.

(1) To make a smaller system for fast screening, first the model of the external system will be reduced in size. The model reduction techniques will be used, and new techniques will be developed to (a) find coherent synchronous

generators in all areas and pack them in equivalent generators for each coherent group, (b) aggregate wind farms, solar power stations, other renewable energy sources, and microgrids in every area, (c) aggregate loads and DERs in distribution networks and make equivalent models of active distribution networks, and (d) make a reduced size equivalent model of the whole external system. This work in progress is covered by two chapters in this book by Adhikarige *et al.*, i.e. Chapters 4 and 5.

(2) A new method in assessing the static PSstr will be worked out, which takes into account the effect of IBRs. Investigation about the impact of control loops of IBRs and their interactions on the system strength is a part of this study. This work in progress is covered by a chapter in this book by Liyanarachchi *et al.*, i.e. Chapter 2.

(3) The critical buses will be determined based on the static values of SCR_rp found in (1.4).

(4) After finding critical buses, full EMT simulations including the details of the external and study systems will be carried out to find dynamic behaviors of IBRs that might be connected at those buses, for various defined scenarios. This step is particularly necessary for the new IBRs that are going to be added based on the connection requests received by the network operators. In this step, it will be determined whether new IBR connections are feasible based on the national electricity law and the relevant policies.

1.2.5 Summary and future directions

In this chapter, the need for finding a new way of determining the level of power system strength (PSstr) in the emerging bulk power systems (BPSs), which will contain various types of inverter-based resources, microgrids, and active distribution networks, was emphasized. A thorough conceptual study of the power system strength and its assessment in IBR-rich power systems was presented. Various definitions of the SCR and its applicability to the new systems with IBRs were reviewed, and some confusions observed in the literature were discussed and clarified.

A summary of an outline for determining PSstr in such BPSs was put forward, which is currently used by the authors in an ongoing research work, and is progressing. In this research, to identify critical buses which will be vulnerable to voltage fluctuations/instability, a fast scheme for the assessment of PSstr is being developed to screen all buses of a BPS, with various RERs in all levels of the system. Once identified, the detailed EMT simulations will be applied to study those critical buses in more detail and with more accuracy. Chapter 3 of this book gives some more insights into this approach.

References

[1] N. Hosseinzadeh, A. Aziz, A. Mahmud, A. Gargoom, and M. Rabbani, "Voltage stability of power systems with renewable-energy inverter-based generators: a review", *MDPI Electronics, Electronics,* 10(2), 115, 2021, Special Issue on Voltage Stability of Microgrids in Power Systems.

[2]   Giles Parkinson, "Flicker fest: Production halted at Australia's biggest wind and solar farm after voltage issues", Renew Economy – Clean Energy News and Analysis, (https://reneweconomy.com.au/flicker-fest-production-halted-at-australias-biggest-wind-and-solar-farm-after-voltage-issues/), 24 June 2022 (Accessed 13 April 2023).

[3]   AEMO, Black System South Australia 28 September 2016 Final Report, Australian Energy Market Operator (AEMO), Melbourne, VIC, Australia, March 2017.

[4]   AEMC, "Mechanisms to enhance resilience in the power system – review of the south Australian black system event", Australian Energy Market Commission (AEMC), Final Report, December 2019.

[5]   C.L. Fortescue, *Method of Symmetrical Co-Ordinates Applied to the Solution of Polyphase Networks*, Transactions of the American Institute of Electrical Engineers, 1918.

[6]   J. Kauferle, R. Mey, and Y. Rogowsky, "H.V.D.C. stations connected to weak A.C. systems", *IEEE Transactions on Power Apparatus And Systems*, Pas-89(7), 1610–1619, 1970.

[7]   IEEE Transmission and Distribution Committee, "IEEE Guide for Planning DC Links Terminating at AC Locations Having Low Short-Circuit Capacities", IEEE Std. 1204-1997 (R2003), June 1997.

[8]   Y.-K. Kim, G.-S. Lee, C.-K. Kim, and S.-I. Moon, "An improved AC system strength measure for evaluation of power stability and temporary over-voltage in hybrid multi-infeed HVDC systems", *IEEE Transactions on Power Delivery*, 37, 638–649, 2021.

[9]   Y. Zhang, S. H. Huang, J. Schmall, J. Conto, J. Billo, and E. Rehman, "Evaluating system strength for large-scale wind plant integration", In *Proceedings of IEEE Power Energy Society General Meeting*, 2014.

[10] S. H. Huang, J. Schmall, J. Conto, Y. Zhang, Y. Li, and J. Billo, "Voltage stability of large-scale wind plants integrated in weak networks: an ERCOT case study", In *Proceedings of IEEE Power Energy Society General Meeting*, 2015.

[11] D. Wu, G. Li, M. Javadi, A.M. Malyscheff, M. Hong, and J.N. Jiang, "Assessing impact of renewable energy integration on system strength using site-dependent short circuit ratio", *IEEE Transactions on Sustainable Energy*, 9(3), 1072–1080, 2018.

[12] N. Choi, B. Lee, D. Kim, and S. Nam, "Interaction boundary determination of renewable energy sources to estimate system strength using the power flow tracing strategy", *Sustainability* 13, 1569, 2021.

[13] CSIRO Australia and EPRI, "CSIRO Australian Research for the GPST, Topic 2 – Stability Tools and Methods", 2021.

[14] AEMO Australia, "Amendments to AEMO instruments for Efficient Management of System Strength Rule", 2022.

[15] A. Gavrilovic, "AC/DC system strength as indicated by short circuit ratios," In *International Conference on AC-DC Power Transmission*, 1991, pp. 27–32.

[16] A. Gavrilovic, "Interaction between ac and dc systems", CIGRE 14.09, 1986.

[17] A. Gavrilovic, P.C.S. Krishnayya, C.O.A. Peixoto, *et al.*, "Some aspects of ac/dc system interconnection", IEEE, Montec, 1986.

[18] A. Gavrilovic, P.C.S. Krishnayya, J.D. Ainsworth, *et al.*, "Interaction between dc and ac systems", CIGRE Symposium, Boston, 1987.

[19] R. Fernandes, S. Achilles, and J. MacDowell, *Report to NERC ERSTF for Composite Short Circuit Ratio (CSCR) Estimation Guideline*, Schenectady, NY: GE Energy Consulting, 2015.

[20] NERC, "Integrating Inverter-Based Resources into Low Short Circuit Strength Systems; North American Electric Reliability Corporation (NERC): Atlanta", GA, USA, 2017.

[21] D. Kim, H. Cho, B. Park, and B. Lee, "Evaluating influence of inverter-based resources on system strength considering inverter interaction level", *MDPI Sustainability (Switzerland)*, 12(8), 3469, 2020, doi:10.3390/su12083469.

[22] L. Yu, H. Sun, S. Xu, B. Zhao, and J. Zhang, "Critical system strength evaluation of a power system with high penetration of renewable energy generations", *CSEE Journal of Power and Energy Systems*, 8(3), 710–720, 2022.

[23] B. Badrzadeh, Z. Emin, S. Goyal, *et al.*, "System strength", *CIGRE Science and Engineering*, 20, 5–26, 2021.

[24] S. Shah, P. Koralewicz, V. Gevorgian, and H. Liu, "Impedance methods for analyzing the stability impacts of inverter-based resources: stability analysis tools for modern power systems", *IEEE Electrification Magazine*, 9, 53–65, 2021.

[25] North American Electric Reliability Corporation (NERC), "Short-Circuit Modeling and System Strength", NERC White Paper, 2018.

Chapter 2

Power system strength assessment with inverter-based resources: challenges and solutions

Lakna Liyanarachchi[1], Nasser Hosseinzadeh[1], Ameen Gargoom[1] and Ehsan Farahani[2]

Abstract

Power system strength is a concept which has been recently defined and assessed in power systems dominated by Inverter Based resources (IBRs). Inverter Based Resources (IBRs) such as solar plants, wind plants, and battery energy storage systems (BESS) have different characteristics to traditional synchronous machines. Unlike rotating machines, which have a natural physical response, IBRs do not behave in the same manner in the power system. This is primarily because they have power electronics interfaces and essentially the behaviors of the IBRs are governed by control loops and control algorithms. So, IBR behaviour at transmission system level needs to be studied in relation to power system strength. This is required to accurately assess their interaction with other system level components . The challenges and solutions involved in power system strength assessment are provided in this chapter.

Acronyms

AC	Alternating current
AEMO	Australian Energy market Operator
BESS	Battery energy storage systems
CSCR	Composite SCR
DC	direct current
EMT	electromagnetic transient
ESCR	Effective SCR
ERCOT	Electric Reliability Council of Texas

[1]School of Engineering, Deakin University, Australia
[2]Australian Energy market Operator (AEMO)

EAC	equal area criterion
FACTS	flexible alternating current transmission system
GFI	grid-following inverter
GFM	grid-forming inverter
GE	General Electric
HVDC	high voltage DC
IBR	inverter-based resource
IILSCR	Inverter interaction level SCR
IGBTs	insulated-gate bipolar transistors
LCC	line-commutated converters
MSCR	Minimum Short Circuit Ratio
NEM	National Electricity Market
PoCs	points of couplings
PLL	phase-locked loop
PSCAD	power systems computer aided design
SCC	short circuit capacity
SCR	short circuit ratio
SCR_rp	short circuit ratio with due consideration to the effect of reactive power injection
SCRIF	SCR with interaction factors
SDSCR	Site-dependent SCR index
SVCs	static var compensators (SVCs)
STATCOMS	static series compensators
VSCs	voltage source converters
WSCR	weighted short circuit ratio

2.1 Introduction

The power system is rapidly transitioning from conventional power systems to accommodate higher penetration of inverter-based resources (IBRs) that are replacing synchronous generators [1]. This power transition previously had power flow in top-down approach from generation to transmission to distribution and to the end users. Disturbances were more predictable, which would make it possible to design robust power systems. However, emerging power systems have bi-directional power flow with the addition of IBRs. This has led to power system operators facing both operational and planning challenges. Over the last decade, with the shift to open markets, the possible sources of disturbances have increased. This means the robustness of the power system is reduced hindering its predictability of the operation [2]. Power system strength is a concept widely spoken about with the inception of IBRs. The power system strength has been related to the voltage stability of the power system through the concept of short circuit capacity (SCC). Recently, the power system strength is defined as the sensitivity of voltage and phase angle to changes in active and reactive power flow from generators including IBRs [2]. Various power system strength assessment methodologies are proposed in the literature for power system strength assessment with IBRs [3]. This chapter provides the theoretical formulations for defining power system strength followed by a comprehensive review of existing power system methodologies capturing the concepts, mathematical formulation, and assumptions along with research gaps. Several case studies to show the performance of short circuit ratio (SCR) index along with proposed solutions to mitigate system strength issues are presented.

2.2 Power system strength with grid-following inverter and grid-forming inverter and its relation to weak grids

Power system strength is mostly spoken in relation to weak areas of the power grids commonly known as weak grids, which can be described in three different perspectives in relation to an IBR-dominated power system as shown in Figure 2.1. It shows that a weak grid can be defined to have high voltage and frequency sensitivity along with low SCC.

High volume of IBRs is often proposed to connect away from load centers and away from synchronous generators to areas rich in wind and sun to benefit from increased power output. However, these points of couplings (PoCs) for IBRs are in electrically weak grid areas where system strength is low as the location is electrically distant from other parts of the power system. Also, IBR developers are selecting sites so that multiple IBRs could be connected nearby. The rapid pace and scale of IBR connections in low system strength areas have led to power system

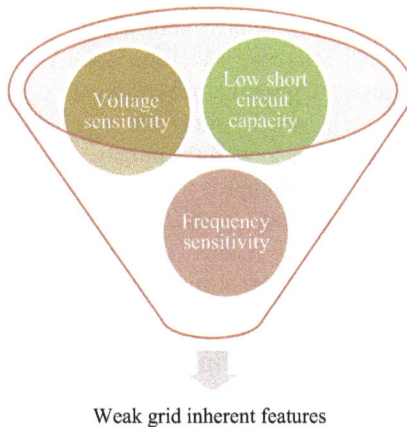

Weak grid inherent features

Figure 2.1   Inherent features of a weak grid

operators facing new technical challenges in maintaining system stability when transmitting power over large distances. Power system operators are required to understand the control interactions between multiple IBRs and interactions between other components at the system level to maintain system security.

A grid-following inverter (GFI) controls the current to the grid at its PoC and is synchronized with the grid voltage through a phase-locked loop (PLL). GFIs become unstable when connected at weak PoCs. A weak PoC is generally represented as a location that is distant from the bulk power system through a long transmission line. The connection of multiple GFIs itself at a weak PoC contributes to the further weakening of the PoC. So, in weak grids, when there is a disturbance or small changes such as changes in load or switching devices, the voltage will be highly sensitive to a small change in current injection leading to a large change in voltage or the rate of change of voltage. Also, the rate of change of frequency could be high in a weak grid. So, in defining a weak grid scenario with GFI behavior, high voltage sensitivity, high-frequency sensitivity, and SCC could be present at the same time or not, depending on the system. When these changes occur in a weak grid, the PLL of the GFI IBR can become unstable leading to control instability. A grid-forming inverter (GFM) controls the grid side voltage and synchronization happens through frequency droop control and stable operation under weak grid conditions is expected [3].

Maintaining the system voltage stability and frequency in weak grid areas is a challenging issue for power system planners and operators with the high uptake of IBRs [1]. The majority of the IBRs that have penetrated the grid to date are GFI due to its simple control structure, mature PLL technology, and as they operate by injection of controlled current. The voltage instability mainly occurred in weak areas with changes in the reactive power. However, control loops of the IBRs specifically that of GFIs further worsen the voltage stability. That is, in weak grids, the PLL of GFIs will be unable to track the voltage reference during a disturbance leading to controller-driven instability. Also, in a weak grid, the IBRs struggle to inject power due to the high Thevenin impedance seen at the PoC. System strength deteriorates in a power system with the inclusion of a high level of IBRs. However, GFM provides fault current unlike GFIs and when connected in weak grid areas having GFIs it can theoretically provide a voltage reference waveform to dampen the voltage oscillations of GFIs. Evaluating the level of system strength at any point of a power system is becoming more vital with the new generation mix which includes more IBRs replacing synchronous generators and also with the attention towards GFIs. However, for operational decisions, this evaluation needs to be done accurately and relatively in a short time. Unfortunately, the increase in accuracy and the reduction of the computation time are contradicting. Therefore, there is a need to find methods for assessing the power system strength in a fast, yet sufficiently accurate way [1].

2.3   Power system strength definitions

In conventional power systems, power system strength has been related to only the SCC, which is also referred to as the fault level at a bus. The voltage strength of a power system is measured as the SCC [2]. In another way, in conventional power systems, power system strength has been associated with the power system impedance and the inertia [2]. The power system impedance consists of the impedances of system-level components such as synchronous generators, transformers, transmission lines, and various loads. As such, when the grid is weak, the impedance will be high. This means that there can be higher voltage variations at buses having lower SCC and being connected to the system with higher equivalent impedance.

System strength at a given bus has been usually determined by:

- number of synchronous machines connected nearby
- number of lines connecting synchronous machines to the network
- equivalent Thevenin impedance

When defining system strength, SCC has a meaning only if it is compared with the connected power or the reactive power of the load [4]. This comparison has led to the SCR index and is discussed in detail in the following section.

Recently, in the context of connecting IBRs to the system, the power system strength has been more broadly evaluated as the sensitivity of the power system to active power and reactive power flow of IBRs. With the inception of IBRs, the power system strength is described as the stiffness of the grid to small changes in load or operation of switching devices [5]. In other words, the sensitivity of a power system to various disturbances determines the system strength [6]. In the National Electricity Market (NEM), power system strength is defined as a concept that can be evaluated by the level of sensitivity of voltage magnitude and its angle to changes in the active and reactive power flow [7].

In a two-bus equivalent with an IBR connected to the receiving end bus $V_{R,i}$ as shown in Figure 2.2, the voltage sensitivity to a small disturbance ΔP and ΔQ can be written as (2.1) and (2.2) and the derivations can be found in Ref. [8]:

$$\frac{\partial V_r}{\partial P_r} = \frac{(x^2 + r^2)P_r + rV_r^2}{V_r\{V_s^2 - 2V_r^2 - 2(rP_r + xQ_r)\}} \tag{2.1}$$

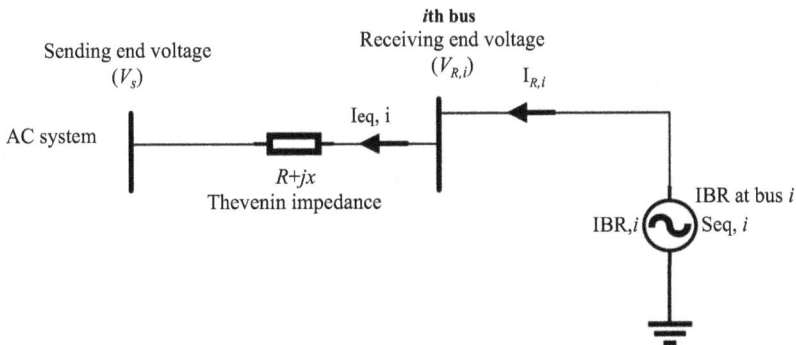

Figure 2.2   Two-bus equivalent with IBR connected to ith bus

$$\frac{\partial V_r}{\partial Q_r} = \frac{(x^2 + r^2)P_r + xV_r^2}{V_r\{V_s^2 - 2V_r^2 - 2(rP_r + xQ_r)\}} \tag{2.2}$$

where V_s is the sending end voltage, P_r is the active power injected by the IBR, Q_r is the reactive power injected by the IBR, and $R + jx$ is the line impedance.

When $x>>r$, then:

$$0 > \frac{\partial V_r}{\partial P_r} >> \frac{\partial V_r}{\partial Q_r} \tag{2.3}$$

The sensitivity of voltage to reactive power flow is much higher than that to P when $x>>>r$ as evident from (2.3).

2.4   System strength metrics

The SCR index is a simple but commonly used indicator all over the world including in the NEM of Australia to assess the power system strength of an IBR bus. To date extended SCR-based indices that claim to include the impacts of multiple IBRs at the vicinity of a point of connection (PoC) are proposed in literature. However, these indices are static indices calculated through load flow parameters and do not account for the unique dynamics of IBRs. The main drawbacks of these indices are that they do not consider the effect of control loops of the IBRs and the dynamics of other power system components in the power system. The below subsection explains the static type system strength indices to date.

2.4.1   SCR index

To assess the system strength, the SCR index is commonly used in IBR-connected systems, mainly due to its simplicity. Earlier, the SCR index was widely used to determine the power system strength in power networks dominated by synchronous generators. When the AC/DC transmission became common in power systems, the SCR has been used to determine the system strength on the AC side due to the interaction from power electronics interfaces [9]. More recently, it is used with modifications to determine the power system strength in IBR-dominated systems specifically for those buses in the grid where IBRs are connected.

The SCC in IBR-dominated systems is calculated by applying a three-phase fault at the point of coupling (PoC) and measuring the fault current injected to the fault from all system-level components, other than any IBR connected to the PoC. The value of SCC is constant at the PoC in the system under study. Then, the SCR at a bus is calculated as the ratio between the SCC in MVA and the rated power (P) in MW of the injected source at that bus. At an IBR-connected bus, the active power of the injected source becomes the rated power of the IBR. Hence, the SCR can be written as in (2.4):

$$SCR = \frac{\text{Short circuit capacity of the PoC without IBR (SCC)}}{\text{Rated power of the IBR connected to the POC }(P)} \tag{2.4}$$

Commonly, an SCR $>> 3$ at a bus is categorized as a strong bus, an SCR value between 2 and 3 is generally considered as a weak bus and SCR$<<2$ is considered a very weak bus as shown by case studies done using average inverter data and on the assumption that inverter is operating at unity power factor [9]. The critical SCR is approximately equal to 2 which represents the border between stable and unstable regions of a rectifier when operating in constant power control and inverter on constant commutation margin control [9]. The SCR ranges provided in Ref. [9] have been commonly used in IBR-dominated systems to distinguish between strong and weak buses; however, the threshold of SCR needs to be used carefully as it will depend on factors such as system characteristics, inverter control technology, and system components. It is well understood that in IBR-dominated systems, lower SCR will indicate that the IBR will run into control issues. However, the critical limitation of using SCR-based screening indices is that it does not give an indication of the exact reason for IBR failure or the exact operating condition at which the IBR fails. Furthermore, an SCR index can only give a higher level of understanding about system strength when a single IBR is connected to the PoC, and its usability deteriorates as more than one IBRs are connected in the vicinity of the PoC.

For a multiple IBR-connected system, the critical SCR has been mathematically calculated using the static voltage stability concept as 2 and is interpreted that Composite SCR (CSCR) will not exceed 2 when IBRs supply reactive power to the power system [10]. This is particularly important as GFM inverters are emerging which provides system strength to the power system. As SCR does not account for IBR reactive power, the critical RESCR has been defined in [9] to account for reactive power from the inverter. Application of the same concept to IBR dominated power system, we propose (2.5) to better define the SCR index to capture the reactive power injection/absorption capability of the IBR as SCR_rp as discussed in detail in Chapter 1.

$$\text{SCR\_rp} = \frac{\text{Short circuit capacity of the PoC with IBR (SCC)}}{P \pm Q} \tag{2.5}$$

Where $+Q$ will be used if reactive power is absorbed by the IBR from the grid and $-Q$ will be used if reactive power is injected into the grid. The proposed index shows that when Q is injected from IBR, then SCR will increase showing that system strength has increased.

Next, definitions and applications of some indices extended from SCR are presented as follows.

2.4.2 Weighted short circuit ratio

The weighted short circuit ratio (WSCR) was derived by the Electric Reliability Council of Texas (ERCOT) [11]. The WSCR calculation takes into account all IBRs in an area and calculates a single index assuming that all IBR are connected at the same PoC.

The WSCR can be written as given in (2.6)–(2.8):

$$\text{WSCR} = \frac{\text{Weighted SCC (WSCC)}}{\text{Total power injected from all IBRs (P)}} \tag{2.6}$$

$$\text{WSCR} = \frac{\left(\sum_i^N SCC_i \times P_{R,i}\right) / \sum_i^N P_{R,i}}{\sum_i^N P_{R,i}} \tag{2.7}$$

$$\text{WSCR} = \frac{\left(\sum_i^N SCC_i \times P_{R,i}\right)}{\left(\sum_i^N P_{R,i}\right)^2} \tag{2.8}$$

where N is the number of IBRs in the region, SCC is the short circuit capacity at the ith bus calculated before IBR connection, and $P_{R,I}$ is the rated power of the IBR in MW.

The ERCOT has derived the WSCR index based on the Panhandle region which has wind plants close to each other and far away from loads. The WSCR index may not be applicable to assess power system strength when the IBRS are dispersed and loads are present in the vicinity. This is because the electrical distance between IBRs will be then significant and it is questionable how to select the number of IBRs for index calculation and also its validity. Further, the load dynamics needs will need to be considered. In the Panhandle region of Texas, the capacities of the wind power plants and the SCC at the buses where they are connected are given in Table 2.1.

The WSCR for the data in Table 2.1 is calculated as shown in (2.9) and (2.10):

$$\text{WSCR} = \frac{(1{,}200 \times 6{,}500) + (1{,}000 \times 8{,}000) + (800 \times 8{,}500) + (2{,}000 \times 7{,}000)}{(1{,}200 + 1{,}000 + 800 + 2{,}000)^2} \tag{2.9}$$

$$\text{WSCR} = 1.464 \tag{2.10}$$

where the WSCR greater than 1.5 indicates a strong system, otherwise, it is considered as a weak system.

The interpretation of the WSCR index will depend on the system characteristics and operating condition under which it is calculated. Thus, considering WSCR > 1.5 to represent a strong region is not applicable to all power systems.

Table 2.1   Panhandle region wind projects (courtesy of [11])

Wind plant	Wind capacity (MW)	Short circuit capacity (MVA)	SCR
A	1,200	6,500	5.42
B	1,000	8,000	8.00
C	800	8,500	10.63
D	2,000	7,000	3.5

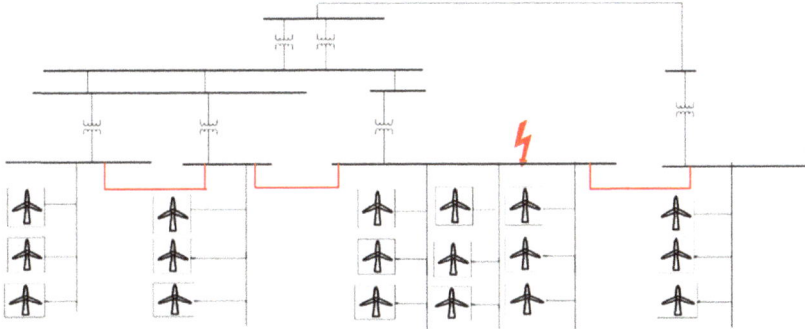

Figure 2.3   Determination of CSCR at the composite bus

2.4.3   Composite SCR

The CSCR has been proposed by General Electric (GE) [12]. The CSCR index is calculated at the composite bus as shown in Figure 2.3.

The CSCR index at the composite bus can be written as shown by (2.11):

$$\text{CSCR} = \frac{\text{SCC at composite bus}}{\text{Total power injected from all IBRs}} \tag{2.11}$$

The CSCR index assumes that all IBRs are close to each other so that these can be tied up and connected to a single bus called the composite bus. The CSCR index is calculated at the composite bus. For CSCR > 2.5, the PoC where IBR is connected is considered as strong and it is interpreted that the IBR connected at this PoC will not result in control instability. For CSCR < 1.7, it indicates that the PoC is a weak one. CSCR < 1 means that there can be control instability of the IBR and remedial actions are required to increase system strength at that PoC.

Both WSCR and CSCR indexes are superior to the SCR index as the interactions of other IBRs are considered in terms of short-circuit capacity. However, both of these indices ignored the real electrical connections in terms of the electric impedance that exists between the interacting IBRs, i.e. WSCR index assumes that all other IBRs are connected at the same PoC, and CSCR assumes all IBRs are connected to a virtual PoC. Further, the given ranges for WSCR and CSCR depend on the power system and the operating conditions at the time it is calculated.

2.4.4   Effective SCR

The effective (ESCR) defined in [9] is provided as an indication of AC/DC system strength as shown in (2.11). The ESCR was defined near HVDC stations where there were HVDC converters of type line-commutated converters (LCC) connected to the grid by a DC link. At steady state, HVDC converters required reactive power to be fed from the grid for the commutation of thyristors. This reactive power was provided by capacitors, filter circuits, and synchronous condensers.

The ESCR has been defined to consider the effect of the capacitor bank connected to the AC grid fed by the HVDC converter [9]. This can be better explained by when there is an event in the power system and voltage at the DC link drops, then more reactive power will be consumed by the HVDC converter. Near voltage collapse point, the reactive power injected from capacitors reaches its maximum level; from that point, any further reduction of voltage needs more reactive power injection from the capacitors, but this is not available. So, in ESCR, this impact of capacitors has been considered as shown by (2.12):

$$\text{ESCR} = \frac{\left(SCC - Q_{capacitors}\right)}{P_{HVDC}} \tag{2.12}$$

where SCC is the SCC (MVA), P_{HVDC} is the rated DC power of the HVDC converter, $Q_{capacitors}$ is the reactive power of all the capacitors in MVar including ac filter capacitors

2.4.5   SCR with interaction factors

The SCR with interaction factors (SCRIF) is based on analyzing interactions of multiple HDVC links in a region. The SCRIF takes into consideration the sharing of system strengths by nearby IBRs [6]. When multiple IBRs are electrically close the system strength is shared by them.

The power system equation for the network in Figure 2.4 can be written as shown by (2.13)

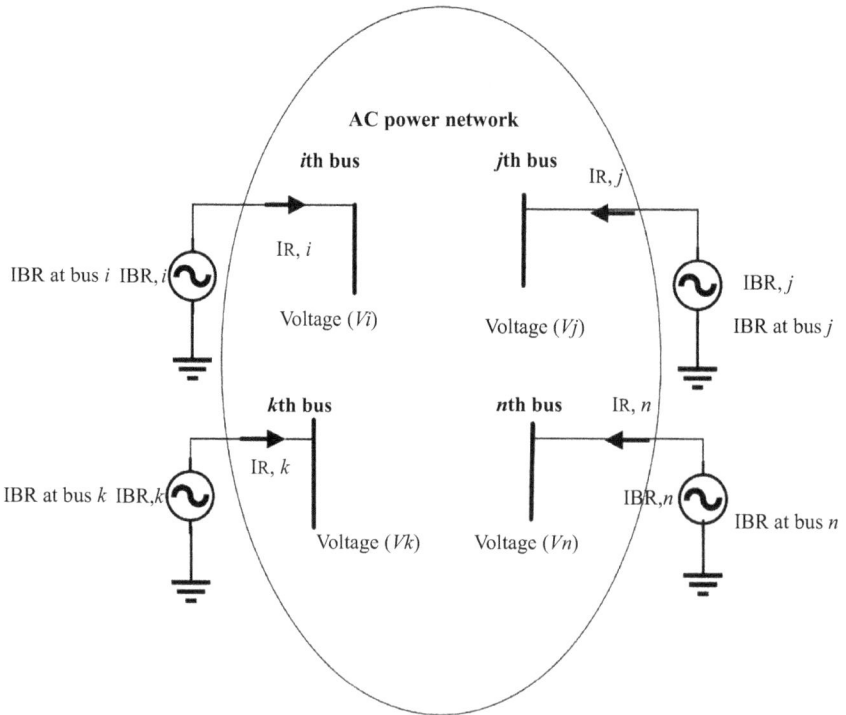

Figure 2.4   Power system with multiple IBRs

$$\begin{bmatrix} V_i \\ V_j \\ V_n \end{bmatrix} = [Z] \begin{bmatrix} I_i \\ I_j \\ I_n \end{bmatrix} \qquad (2.13)$$

Assuming the current at the *i*th bus changes due to a small change of voltage at the *j*th bus, (2.14) and (2.15) are obtained

$$\Delta V_j = Z_{ji}\Delta I_i \qquad (2.14)$$

$$\Delta V_i = Z_{ii}\Delta I_i \qquad (2.15)$$

The change in voltage ΔV_i at a bus connected with an IBR to a change in voltage ΔV_j at another IBR-connected bus is written as the wind plant interaction factor (WPIF). The WPIF is shown in (2.16):

$$\mathrm{WPIF}_{ij} = \frac{\Delta V_i}{\Delta V_j} = \frac{Z_{ji}}{Z_{ii}} \qquad (2.16)$$

The WPIF will be close to zero when the IBRs are electrically far and close to 1 when IBRs are close to each other. The SCRIF index is shown by (2.17):

$$\mathrm{SCRIF}_i = \frac{S_i}{P_{WF,i} + \sum_j \left(WPIF_{ji} \times P_{WFj} \right)} \qquad (2.17)$$

where S_i is the SCC at the *i*th bus, $P_{WF,i}$ is the power injected by the IBR at the *i*th bus, $P_{WF,j}$ is the power injected by the IBR at the *j*th bus. SCRIF index can capture voltage changes when reactive power is injected or absorbed at the IBR buses. However, using this index in an online security assessment platform would be quite challenging specifically when deciding on the type of actions the power system operator needs to take when the index fluctuates.

2.4.6 Site-dependent SCR (SDSCR) index

The site-dependent SCR (SDSCR) index considers the real interactions of other IBRs by including their electrical distance with respect to the reference IBR [13]. The electrical distance is considered by taking the impedance matrix into the analysis. A higher value recorded for Thevenin impedance will mean that the SDSCR ratio is lower which further means that system strength is lower. According to the studies done with the SDSCR index having a value higher than 3 indicates a strong system and if its value is between 2 and 3, the system will be identified as a weak system while the SDSCR lower than 2 represents a very weak system. At the voltage collapse point, the SDSCR is mathematically equated to 1, meaning the system will result in voltage instability [13].

The network equation of a power system having multiple IBRs and synchronous generators as shown in Figure 2.5 can be represented by equation (2.17):

$$\begin{bmatrix} V_G \\ V_R \end{bmatrix} = \begin{bmatrix} Z_{GG} & Z_{GR} \\ Z_{RG} & Z_{RR} \end{bmatrix} \begin{bmatrix} I_G \\ i_R \end{bmatrix} \qquad (2.18)$$

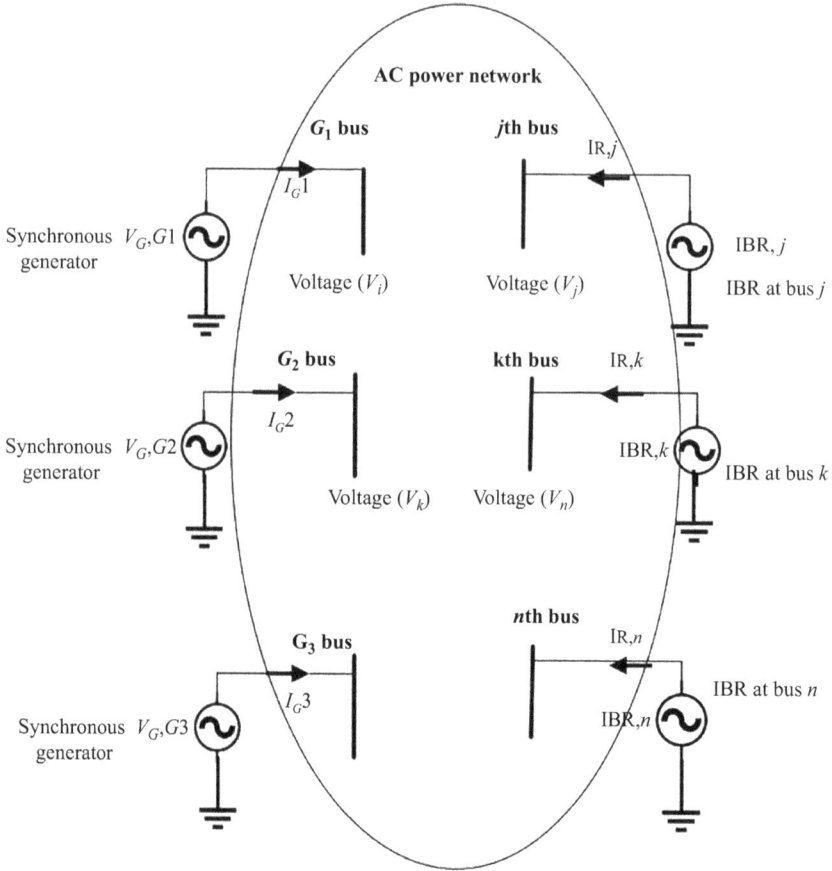

Figure 2.5   AC power network with multiple IBRs and synchronous generators

where G represents the buses with synchronous generators, R represents the buses with IBRS, I_G is the current injected into the network by synchronous generators, I_R is the current injected into the network by IBRs, V_G is the voltage at the buses connected with synchronous generators, and V_R is the voltage at the buses connected with IBRs. The bus impedance matrix is represented by matrices Z_{GG}, Z_{GR}, Z_{RG}, Z_{RR}.

The voltage stability boundary of the ith bus is given by (2.19) with reference to Figure 2.6.

$$V_{R,i} = \sum_{k \in G} Z_{RG,i.k} I_{G.k} + Z_{RR,ii} \sum_{j \in R} \frac{Z_{RR,ij}}{Z_{RR,ii}} I_{R,j} \tag{2.19}$$

where $V_{R,i}$ is the voltage of the ith bus which is an element of V_R vector, $I_{R,j}$ is the current injected by the IBR at the jth bus which is an element of $I_{G,k}$ vector, $Z_{RG,ik}$ is (i,k)th element of the impedance matrix, $Z_{RR,ij}$ is the (i,j)th element of Z_{RR} matrix.

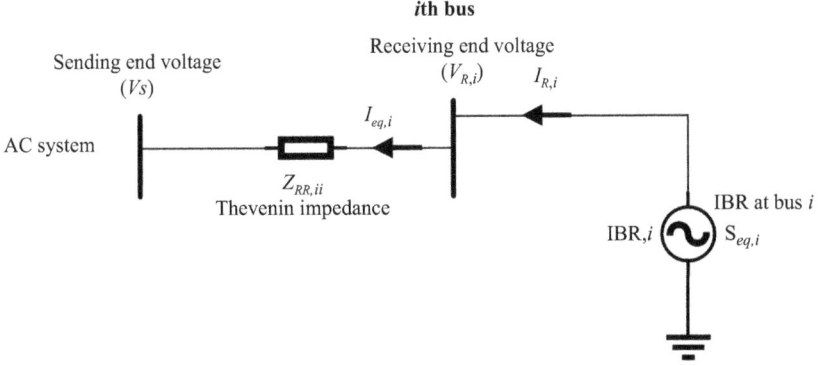

Figure 2.6 Two bus equivalent of the power system with an IBR connected

The power flow equation of the system is given by (2.20):

$$\frac{V_{R,i} - V_s}{Z_{RR,ii}} = I_{eq,i} = \left(\frac{S_{eq,i}}{V_{R,i}}\right)^* \tag{2.20}$$

The complex power at the *i*th bus after considering other IBR interactions can be written as in (2.21)

$$S_{eq,i} = V_{R,i}\left(I^*_{R,i} + \sum_{j \in R, j \neq i} \frac{Z^*_{RR,ij}}{Z^*_{RR,ii}} I^*_{R,j}\right) \tag{2.21}$$

The SDSCR index at the *i*th bus is given by (2.22).

$$SDSCR_i = \frac{|V_{R,i}|^2}{\left(P_{R,i} + \sum_{j \in R, j \neq i} P_{R,j}\omega_{ij}\right)|Z_{RR,ii}|} \tag{2.22}$$

$$\omega_{ij} = \frac{Z_{RR,ij}}{Z_{RR,ii}} \times \left(\frac{V_{R,i}}{V_{R,j}}\right)^* \tag{2.23}$$

where $V_{R,i}$ is the voltage of the *i*th bus, V_s is the sending end voltage, $Z_{RR,ii}$ is the Thevenin impedance seen at bus *i*, $I_{eq,i}$ is the current injected by the IBR, and $S_{eq,i}$ is the complex power injected by the IBR.

2.4.7 Inverter interaction level SCR (IILSCR)

The interactions of multiple IBRs to a reference IBR connected at a PoC bus is given by the IILSCR index [15]. The sum of the power received by other IBRs at the reference Bus 'i' in Figure 2.7 is considered in the IILSCR index to account for the interactions of the other IBRs to the reference IBR.

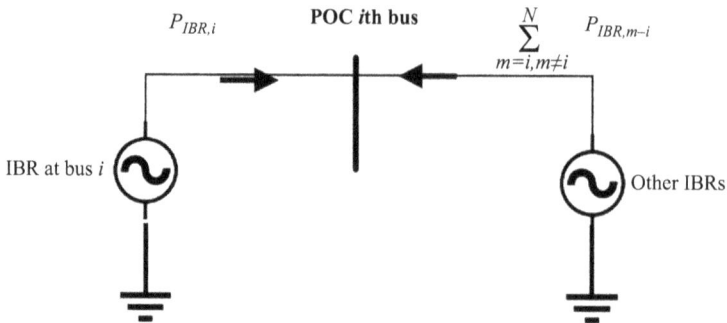

Figure 2.7   IBR connected at bus i *considering the influence of other IBRs*

The IILSCR is given by equation (2.24)

$$\text{IILSCR}_i = \frac{SCC_i}{P_{IBR_i} + \sum_{m=1,m\neq i}^{N} P_{IBRm-i}} \tag{2.24}$$

where SCC_i is the SCC in MVA at the ith bus before IBR connection, $P_{IBR,i}$ is the injected power of the IBR to the ith bus, and P_{IBRmi} is the power injected by other nearby IBRs.

The IILSCR index takes into account the electrical distance between IBR-connected buses by tracing the active power through power flow tracing methodology. Case studies done with the IILSCR index shows that if index falls below 3, then the system is considered to be weak as in the case with an SCR index. However, the IILSCR index suffers the drawbacks of the SCR index itself.

2.4.8   Attributes of power system strength assessment methodologies

System strength index	Attributes
SCR	Simple and commonly used index
	SCR metric quantifies system strength only for a single proposed IBR. The interactions between other IBRs, synchronous condensers, and flexible alternating current transmission system (FACTS) devices are not considered.
WSCR	Takes into account the interaction of multiple IBRs. WSCR index calculates a single index for a group of IBRs electrically near by assuming all other IBRs are connected at the same PoC
CSCR	Takes into account the interaction of multiple IBRs nearby by assuming that the IBRs are connected at a virtual PoC generally known as the composite bus.
SCRIF	Takes into account the electrical distance between multiple IBRs. The voltage change due to reactive power injection from FACTS devices at other buses to the bus where IBR is connected is considered

(Continues)

(Continued)

System strength index	Attributes
SDSCR	Takes into account the electrical distance in terms of electrical impedance between multiple IBRs
IILSCR	Takes into account the electrical distance in terms of electrical impedance between multiple IBRs by tracing the active power flow

2.5 Impact of power system components on power system strength

2.5.1 Impact of phase-locked loops on the system strength

The majority of the IBRs are type IV-connected wind farms and grid-connected photovoltaic systems which are voltage source converters (VSCs). There is full capability for its controller to control the active and reactive current that it injects into the grid in an independent manner leading to fast control of active and reactive power. However, the current injection has to be within the limits to protect the power electronic switches of the inverter which are commonly insulated-gate bipolar transistors (IGBTs). Overall, in many cases, there will be fast, yet accurate control of power from IBRs which is advantageous [2].

An IBR is connected to the grid through the inverter and AC filter. The control loops of the IBRs have three types, that is, phase locked loop (PLL), inner-current control loop, and high-level control loop. The interfaces and control loops of IBRs are shown in Figure 2.8.

The IBR resource such as solar, wind, and battery banks connect to the DC bus represented as V_{dc}. The DC is converted through the inverter to an AC which is fed to the grid through the AC filter. The PLL enables the IBR to be operated in synchronism with the grid. The PLL tracks the voltage angle of the grid and provides this angle as the reference to be used for the high-level control loop and the current control loop. Active power and reactive power control, fault ride through control, and frequency response are done by the high-level control of the IBR. Out of these controls, fault ride through control is faster than the other controls. All other controls except the fault ride through are slower than the inner current control loop. IBRs ability to respond and successfully ride through a fault depends on the stability of these controls [2].

The key challenges in system dynamics due to IBRs are as follows [2]:

1. IBRs provide limited fault current contribution typically in the range of 0–1.5 pu. Doubly fed induction generator wind turbines which are known also as type 3 wind turbines are capable of supplying more fault current due to the direct coupling of the stator onto the grid.

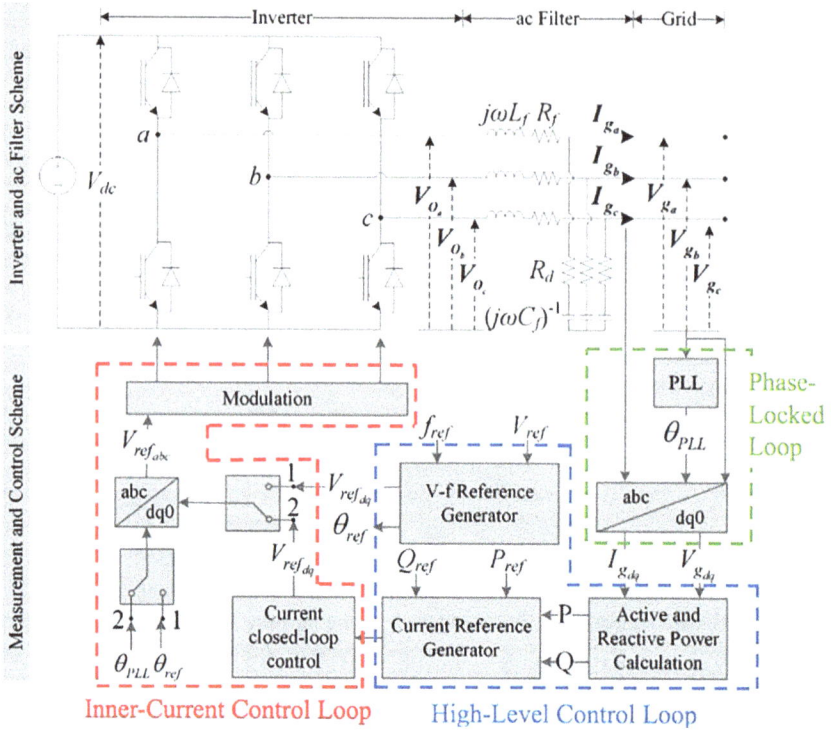

Figure 2.8   IBR inverter and controls (courtesy of [16])

2. The recovery of an IBR after a disturbance majorly depends on the PLL and inner current control loops. In weak grids, due to low SCR, these two loops tend to become oscillatory. This is due to the inability of the PLL to synchronize with the grid voltage or because of high gains in the PLL loop or inner current control loop.

If a disturbance happens when a wind farm is delivering a large amount of power, the SCR will further be reduced at the PoC due to a loss in the transmission line impedance. This causes PLL oscillations to increase. The PLL parameters have a direct impact on oscillations and these are significant as they have high participation factor than stator oscillations and mechanical oscillations [17]. If the proportional gain (Kp) of PLL is increased, it will reduce the damping ratio of the PLL oscillation mode. Through the modal analysis, the PLL mode is identified to have a greater impact in terms of the oscillation modes observed in grid-connected wind farms. Thus, PLL mode oscillations are responsible for instability issues with wind farms apart from the low system strength at the PoC. The importance of considering the gain of PLL is highlighted in [17].

The design of IBRs by different vendors and the dynamics of the PLL and inner current control loop determine the dynamic behavior of an IBR. The stability of VSCs can be studied as small signal stability and large signal stability also known as transient stability. Small signal stability studies determine whether the VSC can remain in synchronism when subjected to small disturbances. Linearization theory has been used to model VSCs for small signal stability studies by considering dynamic impacts of the PLL, active and reactive power control loop, current control loop, voltage control loop. For small signal stability studies, eigen-value analysis and impedance type analysis are used.

Large signal stability studies take into account the effects of large disturbances such as grid faults to determine whether there exists a stable equilibrium point in faulty or normal conditions. The system is said to be transient unstable if there are no stable equilibrium points or if there is insufficient damping to reach an equilibrium point. For transient stability studies involving the PLL, it is assumed that the VSC is an ideal current source. This assumption is taken as the PLL loop dynamics is slower than the current control loop of the VSC. The PLL is transiently stable means that the synchronism remains between the VSC and grid. The transient stability with IBRs is comprehensively analyzed through the equal area criterion (EAC) but this assessment neglects damping, that is, Kp = 0 as suggested in [18]. So, this type of study fails to provide a realistic transient stability study of the PLL. To overcome this gap, the phase portrait method has been used for transient stability analysis of VSCs [14]. This study has concluded that the higher damping ratio and lower settling time of the PLL in a high-voltage transmission grid indicate that it is stable. In a low-voltage distribution grid, the settling time has no effect on the PLL transient stability. However, the drawback of this type of non-linear stability analysis is that it does not introduce a generalized rule. To overcome this gap, recently, Lyaponov's direct method has been incorporated for non-linear stability analysis of the PLL [19]. This method takes into account the damping of the system and the current controller is assumed to be ideal as generally current controller dynamics are faster than PLL dynamics. However, this type of method is not suitable when the current controller dynamics are slower and it is required they should be considered in the stability analysis.

2.5.2 *Impact of flexible alternating current transmission system devices on the power system strength*

FACTS devices such as static var compensators (SVCs) and static series compensators (STATCOMS) provide reactive power and can assist in controlling voltage fluctuations in weak grid areas. SVCs are built using the thyristor technology and used for grid voltage regulation and are modeled as a constant impedance in the steady state. STATCOMS are VSC type so its fault current is limited and modeled as a constant current component under fault condition. Both SVCs and STATCOMs are not included in the SCR calculation; however, as SVCs or STATCOMS are installed along with PV or wind farms, the system strength at the PoC will be increased with the connection of FACTs devices. However, FACTS devices provide no added inertia to the power system and have fast-acting control loops that can interact with IBR control loops. In the TransGrid in 2017, one such case has been evident at the Broken Hill substation. The Broken Hill area is a weak grid area where there are solar farms but there are no synchronous generators until Wagga. The transmission line from the Broken hill substation to Buronga is around 250 km and then from Buronga the transmission line is around 400 km. During a planned outage of the Buronga and Redcliff transmission lines, sustained voltage oscillations have been observed due to interactions between the solar farm and the SVC. To mitigate this issue, the SVC has been put into manual mode. The gap in research in this area is that although there are practical cases where SVCs have interacted with IBR control loops, there is no system strength index or algorithm which can detect this behavior and alarm the power system operator in real time [2]. In the preliminary system strength assessment done by AEMO, FACTS devices are not considered for fault level calculations and thus are not reflected in the Minimum Short Circuit Ratio (MSCR) calculation.

In the West Murray region of the National Electricity Market (NEM), there are SVCs along with high penetration of IBRs. There have been sustained post-disturbance voltage oscillations observed in this region with a frequency between 7 and 10 Hz. To identify the key contributors to these oscillations, the electromagnetic transient (EMT) model of the West Murray region has been simulated with and without the SVC. However, it has been identified that when the SVC was taken out of service, the voltage oscillations further increased. So, it has been concluded that the SVC is not contributing to the undesirable voltage oscillations, but some IBRs are positive contributors whereas some IBRs have no effect [2].

2.5.3   Impact of synchronous condensers on the system strength

Synchronous condensers enhance the system strength in weak grids by providing the SCC and inertia. However, due to the high installation cost of synchronous condensers, power system planners and operators should not heavily depend on them as the only solution to improve the system strength in weak grids. Apart from that, in an IBR-dominated system, when a fault happens and clears, synchronous condensers have created inter-area oscillations if there is no damping support [2].

In 2018, two synchronous condensers, each of 175 MVAR, were placed in the Panhandle area of the ERCOT which is highly dominated by IBRs. Dynamic stability studies done in 2019 by the ERCOT indicated 1.8 Hz oscillations. The low damping was observed because the Panhandle area is electrically far away from synchronous generators and the loads are present.

In the South Australian power system, which is highly dominated by IBRs, efforts have been made to provide damping support by using synchronous condensers fitted with power system stabilizers (PSSs) and flywheels [2]. When connecting a synchronous generator to a location with high IBR and few synchronous generators, a system strength study needs to be performed to observe the behaviors of synchronous condensers with the new connection. This will indicate whether adding the synchronous condenser will cause further inter-area oscillations or not.

2.6 Applicability of SCR index: case study

2.6.1 EMT simulations on SCR index

We have done an EMT simulation in PSCAD to study the impact of a type IV wind power plant on the system strength by varying the SCR and applying a three-phase-to-ground fault at 5s and cleared at 5.1s as shown in Figure 2.9, in which an available model of type IV wind farm in PSCAD is used.

2.6.1.1 Case 1: strong system: SCR close to 10

The SCC for a 33 kV system with the equivalent Thevenin impedance of 1 Ω can be calculated as:

$$\text{SCC} = \frac{(\text{Pre-fault grid voltage})^2}{\text{Thevenin impedance}} = \frac{V^2}{Zth} = \frac{33^2 \text{ kV}}{1 \Omega} = 1089 \text{ MVA} \qquad (2.25)$$

If the rated power of the wind power plant is 100 MW, the SCR can be calculated as:

$$SCR = \frac{1,089 \text{ MVA}}{100 \text{ MW}} = 10.89, \text{ represents a strong system} \qquad (2.26)$$

The voltage, active power, and reactive power waveforms for case 1 are shown in Figure 2.10.

2.6.1.2 Case 2: strong system: SCR 3

The SCC for a 33 kV system with the equivalent Thevenin impedance of 3.63 Ω, the SCC can be calculated as:

$$\text{SCC} = \frac{(\text{Pre-fault grid voltage})^2}{\text{Thevenin impedance}} = \frac{V^2}{Zth} = \frac{33^2 \text{ kV}}{3.63 \Omega} = 300 \text{ MVA} \qquad (2.27)$$

For a similar rated power, the SCR can be calculated as:

$$SCR = \frac{300 \quad \text{MVA}}{100 \quad \text{MW}} = 3, \text{ represents a strong system} \qquad (2.28)$$

The voltage, active power, and reactive power waveforms for case 2 are shown in Figure 2.11.

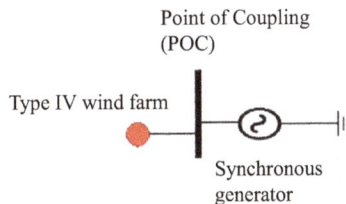

Point of Coupling (POC)

Type IV wind farm

Synchronous generator

Figure 2.9   Wind power plant connected with a synchronous generator at the PoC

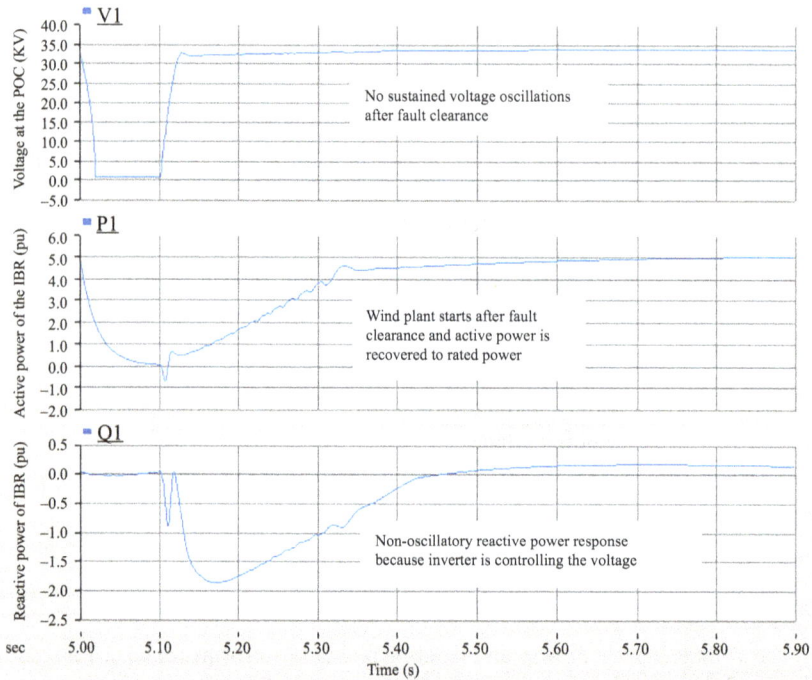

Figure 2.10   Case 1: voltage (N_{rms}), active power (P), and reactive power (Q) for SCR 10

An SVC was connected at the PoC and the waveforms after fault clearance are shown in Figure 2.12. It is evident that as both the SVC and the IBR are attempting to control the voltage at the PoC, there are sustained oscillations seen in voltage, active power, and reactive power waveforms.

2.6.1.3   Case 3: weak system: SCR close to 2

The SCC for a 33 kV system with the equivalent Thevenin impedance of 5 Ω, the SCC can be calculated as:

$$\text{SCC} = \frac{(\text{Pre-fault grid voltage})^2}{\text{Thevenin impedance}} = \frac{V^2}{Zth} = \frac{33^2 \text{ kV}}{5 \text{ }\Omega} = 218 \text{ MVA} \qquad (2.29)$$

For a similar rated power, the SCR can be calculated as:

$$\text{SCR} = \frac{218 \text{ MVA}}{100 \text{ MW}} = 2.1, \text{ represents a weak system} \qquad (2.30)$$

The voltage, active power, and reactive power waveforms for case 3 are shown in Figure 2.13.

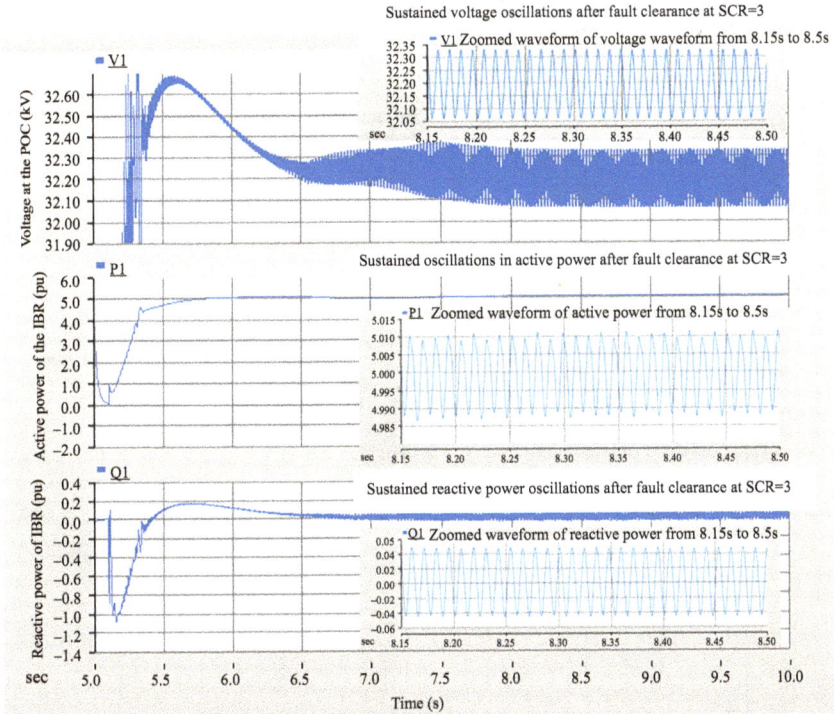

Figure 2.11   Case 2: voltage (V$_{rms}$), active power (P), and reactive power (Q) for SCR 3

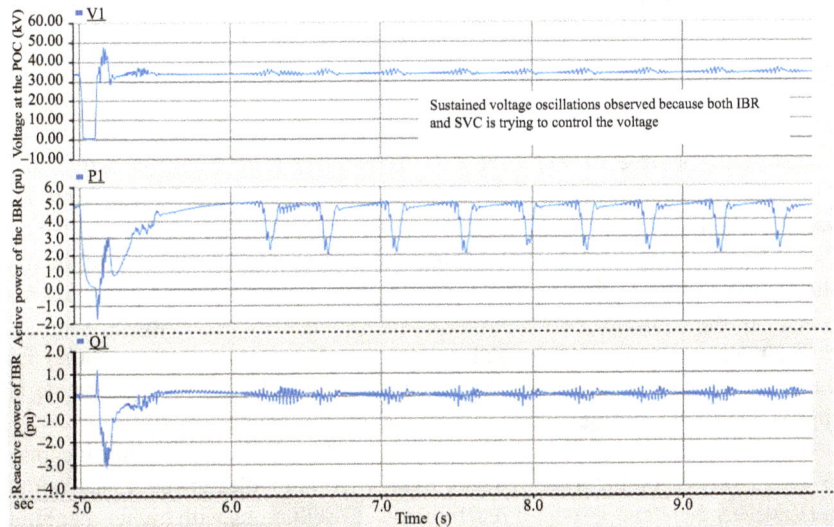

Figure 2.12   Case 2: voltage (V$_{rms}$), active power (P), and reactive power (Q) for SCR 3 with the addition of SVC

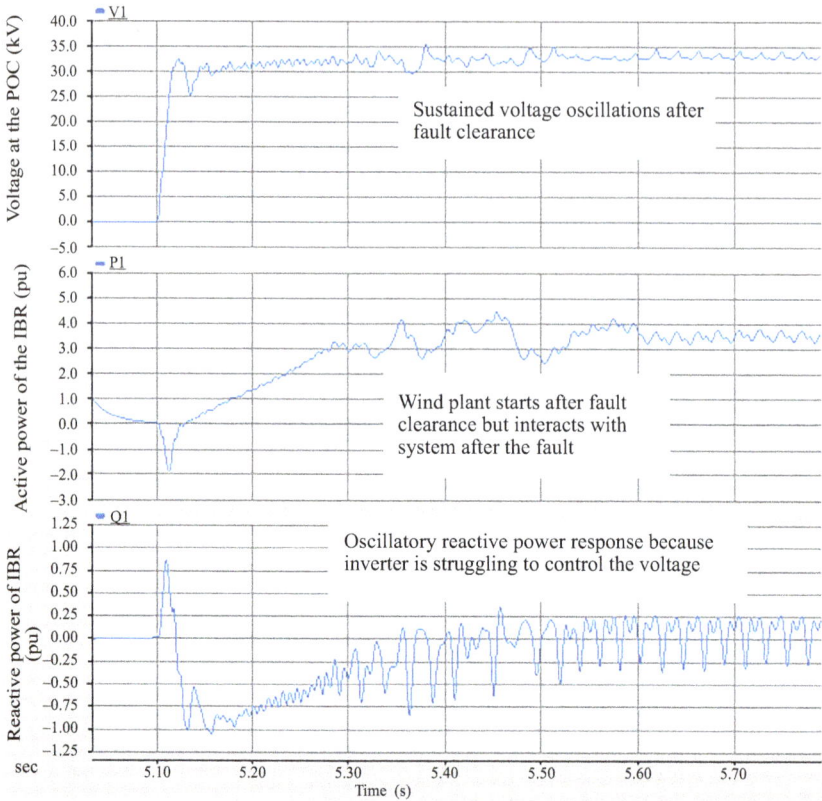

Figure 2.13   Case 3: voltage (V_{rms}), active power (P), and reactive power(Q) for SCR 2

2.6.2   Discussion of simulation results

For case 1, it is observed from Figure 2.10 that at the fault inception, the active power of the IBR is reduced due to activation of fault ride through mode. After fault clearance, the IBR active power ramps up and reaches its rated power. The reactive power support provided by the IBR is supporting the voltage to ramp up to the rated voltage while the active power ramps up to the rated power. As this case represents a strong grid, there are no post-disturbance voltage fluctuations. However, soon after the disturbance, the active power reduces to zero. There is a period soon after fault clearance that the PLL needs to re-synchronize. This behavior is unacceptable and happens due to PLL realignment and giving a temporary angle error.

In case 2, the SCR is 3, and as observed in Figure 2.11, at the fault inception, the active power of the IBR reduces and, after fault clearance, it has ramped up. However, there are post-fault reactive power oscillations with higher magnitude occurring from around 6.5s. So, there are voltage fluctuations that have resulted in active power oscillations.

In case 3, according to the waveforms in Figure 2.13, at the fault inception, the active power of the IBR reduces and, after fault clearance, starts to ramp up. However, as the Thevenin impedance seen by the IBR is larger, the IBR struggles to provide reactive power for voltage support, so the voltage drops. This is seen from the oscillatory reactive power response of the IBR. So, the IBR re-enters fault ride through mode. So, active power is reduced and voltage recovers. However, as this cycle is repetitive, sustained voltage oscillations are observed in the post-disturbance voltage. We may conclude that the SCR index does not give an indication of whether a given IBR will lead to control instability leading to post-disturbance sustained oscillations. Moreover, an SCR index is not able to quantify the magnitude of the voltage oscillations that results from lower SCR.

2.7   Research gaps and new research directions

Existing system strength indices discussed in the literature are derived through the steady-state voltage stability analysis framework considering an SCR index. Efforts have been made to extend SCR index to capture the interactions between multiple IBRs. However, these SCR-extended indices do not account for control interactions. It is required to evaluate the power system strength at a selected location where an IBR is proposed to connect for various dynamic grid conditions. In the vicinity of that location, there will be several components such as other IBRs, FACTS devices, active loads, synchronous condensers, synchronous generators, conventional loads including induction motors, and interconnections. These make the available indices not sufficient to provide the actual information regarding the system strength with IBRs.

SCR index which is commonly used to classify system strength at the PoC of IBRs does not capture the behavior of IBR control loops. If SCR falls less than 2 at a PoC where an IBR is connected, there is a theoretical assumption that IBR performance would be affected by the instability of the PLL and inner current control loops leading to control instability. So, this IBR would not be allowed to connect to the grid. However, the actual system strength at which this instability would occur depends on the IBRs provided by different manufacturers and also on the system operating conditions. Depending on the system conditions, this instability could be prevented if the actions from controls of the IBRs were reduced. However, the system strength at which this should be done and how to select the IBRs, which cause these oscillations, need to be determined. To date, studies done to determine the relationship between voltage stability and system strength indices with IBRs have not extensively incorporated the behavior of fast-acting control loops of IBRs in relation to the assessment of system strength.

References

[1]   L. Liyanarachchi, N. Hosseinzadeh, A. Mahmud, and A. Gargoom, "Challenges in power system strength assessment with inverter-based

resources," In *2021 3rd International Conference on Smart Power & Internet Energy Systems (SPIES)*, Shanghai, China, 2021, pp. 158–163, doi: 10.1109/SPIES52282.2021.9633775.

[2] L. Liyanarachchi, N. Hosseinzadeh, A. Mahmud, A. Gargoom, and E. M. Farahani, "Contingency ranking selection using static security performance indices in future grids," In *2020 Australasian Universities Power Engineering Conference (AUPEC)*, Hobart, Australia, 2020, pp. 1–6.

[3] L. Liyanarachchi, N. Hosseinzadeh, A. Gargoom, and E. M. Farahani, "A new index for the assessment of power system strength with inverter based resources in the presence of static var compensators (SVCs)," *Sustainable and Technology Assessments Journal*, 60, 103460, pp. 1–14, 2023.

[4] A. Gavrilovic, "AC/DC system strength as indicated by short circuit ratios," In *International Conference on AC DC Power Transmission*, pp. 27–32, 1991.

[5] NERC Reliability Guideline, BPS-Connected Inverter-Based Resource Performance, Atlanta, GA: North American Electric Reliability Corporation (NERC), 2018.

[6] NERC, *Integrating Inverter-Based Resources into Low Short Circuit Strength Systems*, Atlanta, GA: North American Electric Reliability Corporation (NERC), 2017.

[7] B. Badrzadeh, Z. Emin, S. Goyal, *et al.*, "System strength," *CIGRE Science and Engineering*, 20, pp. 5–26 2021.

[8] Y. Hase, *Handbook of Power System Engineering*, New York, NY: Wiley, 2007.

[9] A. Gavrilovic, "AC/DC system strength as indicated by short circuit ratios," In *International Conference on AC DC Power Transmission*, pp. 27–32, 1991.

[10] L. Yu, H. Sun, S. Xu, B. Zhao, and J. Zhang, "A critical system strength evaluation of a power system with high penetration of renewable energy generations," *CSEE Journal of Power and Energy Systems*, 8(3), 710–720, 2022, doi:10.17775/CSEEJPES.2021.03020.

[11] S. H. Huang, J. Schmall, J. Conto, Y. Zhang, Y. Li, and J. Billo, "Voltage stability of large-scale wind plants integrated in weak networks: An ERCOT case study," In *Proceedings of IEEE Power Energy Society General Meeting*, pp. 1–5, 2015.

[12] R. Fernandes, S. Achilles, and J. MacDowell, *Report to NERC ERSTF for Composite Short Circuit Ratio (CSCR) Estimation Guideline*, Schenectady, NY: GE Energy Consulting, 2015.

[13] D. Wu, G. Li, M. Javadi, A.M. Malyscheff, M. Hong, and J. N. Jiang, "Assessing impact of renewable energy integration on system strength using site-dependent short circuit ratio", *IEEE Transactions on Sustainable Energy,* 9(3), 1072–1080, 2018.

[14] H. Wu and X. Wang, "Transient stability impact of the phase-locked loop on grid-connected voltage source converters," In *International Power Electronics Conference (IPEC-Niigata 2018 – ECCE Asia), 2018*, 2673–2680, 2018.

[15]　D. Kim, H. Cho, B. Park, and B. Lee, "Evaluating influence of inverter-based resources on system strength considering inverter interaction level," *Sustainability (Switzerland)*,12(8), 3469, 2020.

[16]　IEEE PES-TR 77, "Stability definitions and characterization of dynamic behavior in systems with high penetration of power electronic interfaced technologies", 2020.

[17]　J. Liu, W. Yao, J. Wen, *et al.*, "Impact of power grid strength and PLL parameters on stability of grid-connected DFIG wind farm," *IEEE Transactions on Sustainable Energy*, 11(1), 545–557, 2020.

[18]　Q. Hu, L. Fu, F. Ma, and F. Ji, "Large signal synchronizing instability of PLL-based VSC connected to weak AC grid," *IEEE Transactions on Power Systems*, 34(4), 3220–3229, 2019.

[19]　M. Z. Mansour, S. P. Me, S. Hadavi, B. Badrazadeh, A. Karimi, and B. Bahrani, "Nonlinear transient stability analysis of phased-locked loop based grid-following voltage source converters using Lyapunov's direct method," *IEEE Journal of Emerging and Selected Topics in Power Electronics*, 10, 2699–2709, 2022, doi:10.1109/JESTPE.2021.3057639.

Chapter 3

Voltage sensitivity-based system strength metric[*]

Heng Wu[1]

Abstract

This chapter analyzes the power transfer limit ($P_{\max}$) of inverter-based resources under the weak grid condition. It is pointed out that the impact of different grid parameters and grid configurations on $P_{\max}$ cannot be fully reflected by the short circuit ratio (SCR) and its extended forms, but can be readily captured by voltage sensitivity ($\partial Q/\partial V$). Hence, $\partial Q/\partial V$ turns out to be a more appropriate metric for the assessment of $P_{\max}$ compared to the SCR. The simulation results are given to demonstrate the theoretical analysis.

3.1 Introduction

Inverter-based resources (IBRs), like solar and wind power plants, are usually located far away from load centers and synchronous generators (SGs) with long transmission lines in between, which form a weak grid at the point of connection. Hence, operating the IBR stably under weak grid conditions while maximizing its output power is one of the main concerns for transmission system operators (TSOs) as well as converter manufacturers [1,2]. The stability and power transfer capability ($P_{\max}$) of IBRs can be assessed by performing EMT simulations, which, however, is very time consuming, especially considering multiple operating scenarios of IBRs. Therefore, it is important to perform an initial screening based on an appropriate metric to narrow down risky scenarios for further EMT studies.

The IBR with grid-following control can be modeled as a PQ source [3], and $P_{\max}$ of the IBR is constrained by the angle stability limit ($\partial P/\partial \theta > 0$) and the voltage stability limit ($\partial Q/\partial V > 0$) [4,5]. Since both $\partial P/\partial \theta$ and $\partial Q/\partial V$ are affected by the grid strength that is quantified by the short circuit ratio (SCR), the SCR becomes a widely adopted metric to assess the $P_{\max}$ of IBRs [3,6–9]. For example,

[*]© [2022] IEEE. Portion of this chapter have been reprinted, with permission, from "Operating wind power plants under weak grid conditions considering voltage stability constraints," *IEEE Transactions on Power Electronics*, vol. 37, no. 12, 2022.
[1]AAU Energy, Aalborg University, Denmark

the SCR of the power grid in the Panhandle area, Texas, USA, would be monitored by the Electric Reliability Council of Texas (ERCOT) in real-time, and the output power of the IBR in the Panhandle area needs to be curtailed if the SCR drops below 1.5 [10]. Except for its basic form, many advanced SCRs, e.g., composite SCR (CSCR) [7], weighted SCR (WSCR) [7], and equivalent SCR (ESCR) [3], have been developed to consider interactions between multiple IBRs. Yet, all these SCR-based metrics neglect other factors that would affect $\partial P/\partial\theta$ and $\partial Q/\partial V$ and, hence, might lead to inaccurate prediction of P_{max} in some scenarios [6]. For example, it is revealed in [11] that $\partial P/\partial\theta$ is also affected by the grid impedance angle (X/R ratio). Hence, different P_{max} of IBRs might be yielded under different X/R ratios, even if the SCR of the grid is the same. Another example given by [12] highlights that P_{max} of IBRs is also affected by the existence of local loads, which, however, cannot be captured in the SCR-based metric either [12].

While there is increasing awareness of the shortcomings of SCR-based P_{max} assessment, a very little attempt can be found in identifying a more appropriate metric for assessing P_{max} of converter-based generation units like IBRs. To fill this void, this chapter proposes a voltage sensitivity metric that is arguably more appropriate for P_{max} assessment. The content of this chapter is based on the authors' own publication in [13]. This chapter first derives the analytical representation of $\partial P/\partial\theta$ and $\partial Q/\partial V$, based on which, it is revealed that not only grid parameters (X/R ratio, voltage level, etc.), but also grid configurations (presence of local loads, synchronous condenser, etc.), would affect the value of $\partial P/\partial\theta$ and $\partial Q/\partial V$. As will be illustrated in this chapter, the SCR-based P_{max} assessment would lead to either too optimistic or too pessimistic predictions in these scenarios. Yet, instead of evaluating both $\partial P/\partial\theta$ and $\partial Q/\partial V$ for a more accurate P_{max} assessment, this chapter reveals that it is generally enough to check $\partial Q/\partial V$ only, as $\partial Q/\partial V$ imposes a stricter constraint on P_{max} than $\partial P/\partial\theta$. Hence, $\partial Q/\partial V$ can be adopted as a more appropriate metric for P_{max} assessment compared to the SCR.

3.2   System description

Figure 3.1(a) shows a typical configuration of the IBR-based power plant (the wind power plant (WPP) is used as an example in this work), where multiple IBRs are connected to the AC system through step-up transformers and cables. By adopting the typical grid-following control, the IBR is synchronized with the power grid by means of the phase-locked loop (PLL). The current control (CC) loop is used to regulate the output current of the IBR (i_o) to follow the current references (i_{dref} and i_{qref}) generated by DC-link voltage control (DVC) and reactive power control (RPC). P_{IBR} and Q_{IBR} are the active and the reactive power output of the IBR, respectively.

Figure 3.2 shows the equivalent circuit of the IBR-grid system, where IBR is modeled as an aggregated PQ source [3]. P_{gen}/Q_{gen} are output active/reactive power of the IBR, respectively, i.e., $P_{gen} = \sum P_{IBR}$ and $Q_{gen} = \sum Q_{IBR}$. $R_{th}+jX_{th}$ represents

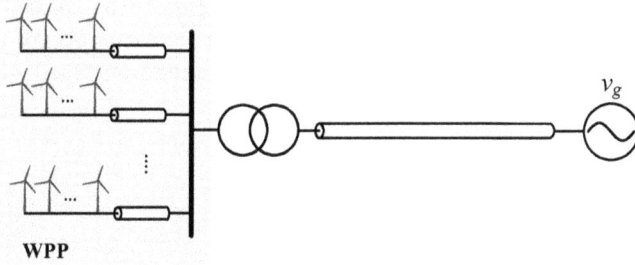

(a)

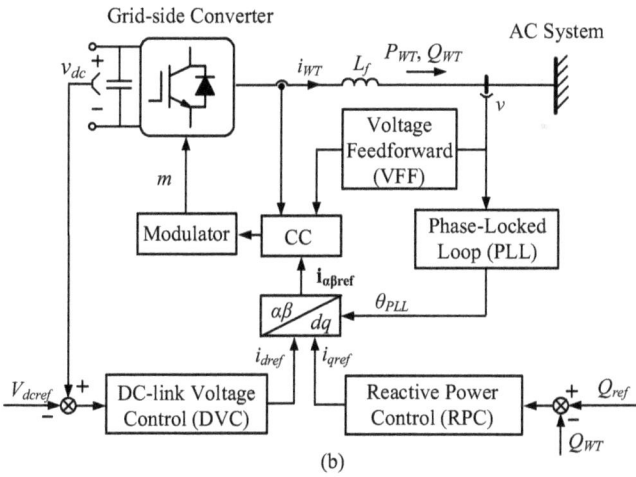

(b)

Figure 3.1   IBR. (a) Typical configuration. (b) Single-line diagram of the grid-following IBR.

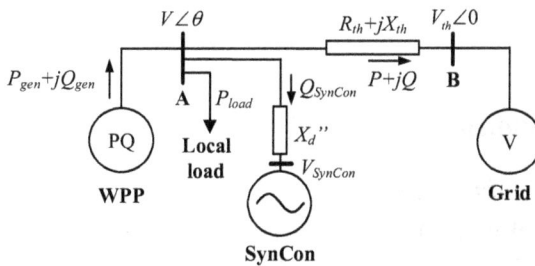

Figure 3.2   Equivalent IBR-grid system

the equivalent line impedance seen from the IBR. V and V_{th} represent RMS values of line-to-line voltages at Bus A and Bus B, respectively. θ denotes the voltage angle differences between these two buses, which is defined as the power angle hereafter. An optional load with active power consumption P_{load} and a synchronous condenser (SynCon) with reactive exchange Q_{SynCon} are also considered at Bus A to illustrate the effect of loading and reactive power compensation of the grid.

The active and reactive power (P and Q) transferred from Bus A to Bus B can be expressed as [14]

$$P = \alpha\left(V^2 - VV_{th}\cos\theta\right) + \beta VV_{th}\sin\theta \tag{3.1}$$

$$Q = \beta\left(V^2 - VV_{th}\cos\theta\right) - \alpha VV_{th}\sin\theta \tag{3.2}$$

where

$$\alpha = \frac{R_{th}}{|Z_{th}|^2} \tag{3.3}$$

$$\beta = \frac{X_{th}}{|Z_{th}|^2} \tag{3.4}$$

$$|Z_{th}| = \sqrt{R_{th}^2 + X_{th}^2} \tag{3.5}$$

By eliminating θ from (3.1) and (3.2), the relationship between P and Q can be further derived as

$$\left(P - \alpha V^2\right)^2 + \left(Q - \beta V^2\right)^2 = \left(\frac{V_{th}V}{|Z_{th}|}\right)^2 \tag{3.6}$$

which forms a circle with center (αV^2, βV^2) and radius $V_{th}V/|Z_{th}|$.

For the inductive transmission line that $X_{th} >> R_{th}$, (3.1) and (3.2) can be simplified as

$$P = \frac{VV_{th}\sin\theta}{X_{th}} \tag{3.7}$$

$$Q = \frac{V^2 - VV_{th}\cos\theta}{X_{th}} \tag{3.8}$$

3.3 Power transfer limit of IBR

3.3.1 *Angle stability limit*

It is well known that the critical power angle at which the angle stability limit is reached can be calculated by solving $\partial P/\partial\theta = 0$ [4]. Based on (3.1), we have

$$\frac{\partial P}{\partial\theta} = VV_{th}(\alpha\sin\theta + \beta\cos\theta) \tag{3.9}$$

Solving $\partial P/\partial\theta = 0$ from (3.9), which yields

$$\theta_{max\_Ang} = 90° + \arctan\frac{R_{th}}{X_{th}} \tag{3.10}$$

Substituting (3.10) into (3.1), which yields

$$P_{\text{max\_Ang}} = \frac{1}{|Z_{th}|} \left(\frac{R_{th}}{|Z_{th}|} V^2 + VV_{th} \right) \tag{3.11}$$

Equation (3.11) illustrates the maximum power transfer capability of the system considering the angle stability limit. Define P_{norm} as the nominal power of the IBR, and then the per-unit (p.u.) representation of (3.11) is given by

$$P_{\text{max\_Ang\_pu}} = \frac{P_{\text{max\_Ang}}}{P_{norm}} = \frac{1}{|Z_{th}|P_{norm}} \left(\frac{R_{th}}{|Z_{th}|} V^2 + VV_{th} \right) \tag{3.12}$$

The SCR is defined as the ratio between the short circuit power from the AC system (S_{SC}) and the rated power of the IBR [3], i.e.:

$$SCR = \frac{S_{SC}}{P_{norm}} \tag{3.13}$$

Without the local SynCon, it is known from Figure 3.2 that S_{SC} can be calculated as

$$S_{SC} = \frac{V_{norm}^2}{|Z_{th}|} \tag{3.14}$$

where V_{norm} represents the nominal voltage. Substituting (3.13) and (3.14) into (3.12), the relationship between $P_{\text{max\_Ang\_pu}}$ and SCR can be derived as

$$P_{\text{max\_Ang\_pu}} = SCR \cdot \left(\frac{R_{th\_pu}}{|Z_{th\_pu}|} V_{pu}^2 + V_{pu} V_{th\_pu} \right) \tag{3.15}$$

It is known from (3.15) that larger R_{th} (lower X/R ratio) and higher V and V_{th} could improve the active power transfer capability. For the special case that $V = V_{th} = 1$ p.u. and $R_{th} = 0$, $P_{\text{max\_Ang\_pu}} = SCR$ is yielded [11].

The above-mentioned conclusions can be visualized by the PQ curve that is plotted based on (3.6) with parameters X/R = 3, SCR = 1.2, $V_{th} = 1$ p.u., as shown in Figure 3.3 (unless otherwise mentioned, only the part with $P \geq 0$ is plotted since IBR can only generate active power). It can be seen that $P_{\text{max\_Ang\_pu}}$ is around 1.6 p.u. with $V = 1$ p.u., which is larger than the SCR due to the presence of R_{th} in the transmission line. Moreover, it can also be observed that $P_{\text{max\_Ang\_pu}}$ can be increased (from around 1.5 p.u. to 1.7 p.u.) by boosting V (from 0.95 p.u. to 1.05 p.u.)

3.3.2 Voltage stability limit

By adopting the grid-following control, the AC voltage of the IBR is maintained by regulating the reactive power of each IBR. Hence, $\partial Q/\partial V > 0$ is mandatory for guaranteeing the voltage stability of the system [5].

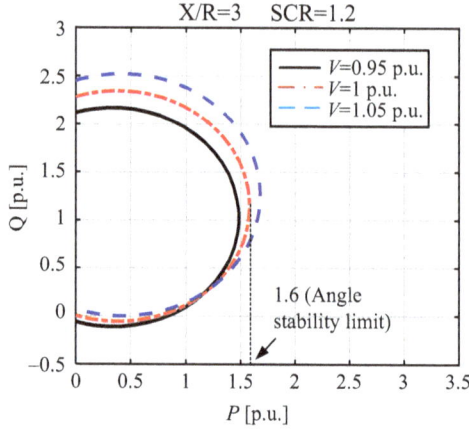

Figure 3.3   P–Q *curve of the system with* X/R = *3,* SCR = *1.2,* V_{th} = *1 p.u.*

Based on (3.6), the relationship between Q and V can be expressed as

$$Q = \beta V^2 - \sqrt{\frac{V_{th}^2 V^2}{|Z_{th}|^2} - (P^2 - 2P\alpha V^2 + \alpha^2 V^4)} \tag{3.16}$$

Differentiating (3.16) with respect to V, which yields [15]

$$\frac{\partial Q}{\partial V} = 2\beta V - \frac{\frac{V_{th}^2 V}{|Z_{th}|^2} + 2P\alpha V - 2\alpha^2 V^3}{\sqrt{\frac{V_{th}^2 V^2}{|Z_{th}|^2} - (P^2 - 2P\alpha V^2 + \alpha^2 V^4)}} \tag{3.17}$$

The p.u. representation of (3.17) can be calculated as (3.18)

$$\frac{\partial Q_{pu}}{\partial V_{pu}} = \frac{1}{|Z_{th\_pu}|}$$

$$\cdot \left(\frac{2X_{th\_pu}}{|Z_{th\_pu}|} V_{pu} - \frac{V_{th\_pu}^2 V_{pu} + 2P_{pu} R_{th\_pu} V_{pu} - 2\frac{R_{th\_pu}^2}{|Z_{th\_pu}|^2} V_{pu}^3}{\sqrt{V_{th\_pu}^2 V_{pu}^2 - \left(P_{pu}^2 |Z_{th\_pu}|^2 - 2P_{pu} R_{th\_pu} V_{pu}^2 + \frac{R_{th\_pu}^2}{|Z_{th\_pu}|^2} V_{pu}^4\right)}} \right) \tag{3.18}$$

If R_{th} is neglected, (3.18) can be simplified as

$$\frac{\partial Q_{pu}}{\partial V_{pu}} = \frac{1}{X_{th\_pu}} \cdot \left(2V_{pu} - \frac{V_{th\_pu}}{\sqrt{1 - \frac{P_{pu}^2 X_{th\_pu}^2}{V_{th\_pu}^2 V_{pu}^2}}} \right) \tag{3.19}$$

Substituting (3.7), (3.13), and (3.14) into (3.19), the relationship between $\partial Q/\partial V$ and power angle θ can be expressed as

$$\frac{\partial Q_{pu}}{\partial V_{pu}} = \frac{1}{X_{th\_pu}} \cdot \left(2V_{pu} - \frac{V_{th\_pu}}{\sqrt{1 - \sin^2\theta}} \right) = SCR \cdot \left(2V_{pu} - \frac{V_{th\_pu}}{\sqrt{1 - \sin^2\theta}} \right) \tag{3.20}$$

For the condition $V = V_{th} = 1$ p.u., it is known from (3.20) that $\partial Q/\partial V = SCR$ when $\theta = 0°$ (i.e., zero active power transfer). Yet, $\partial Q/\partial V$ is reduced with the increased θ (i.e., increased active power transfer). A small $\partial Q/\partial V$ implies that a small amount of reactive power injection would result in a large voltage change at the point of connection of the IBR, which bring challenges to the reactive power control. Moreover, this fast-changing voltage also imposes another challenge to the accurate phase angle tracking of the PLL.

The critical power angle where the voltage stability limit is reached ($\partial Q/\partial V = 0$) can be calculated as

$$\frac{\partial Q_{pu}}{\partial V_{pu}} = 0 \Rightarrow \theta_{\max\_Vol} = \arcsin\sqrt{1 - \left(\frac{V_{th\_pu}}{2V_{pu}} \right)^2} \tag{3.21}$$

It can be calculated from (3.21) that $\theta_{\max\_Vol} = 60°$ when $V = V_{th} = 1$ p.u., which is smaller than the $90°$ power angle limit specified by (3.10). Hence, the voltage stability requirement imposes a stricter constraint on the power transfer capability of the IBR than the angle stability requirement. This conclusion can be visualized by comparing Figures 3.3 and 3.4(a) that is plotted based on (3.18), i.e., with $X/R = 3$, $SCR = 1.2$, $V = V_{th} = 1$ p.u., the $P_{\max}$ of the IBR considering the angle stability limit is around 1.6 p.u. (see red dashed line in Figure 3.3), but is reduced to around 1.2. p.u. when the voltage stability limit is considered, as shown by the black solid line in Figure 3.4(a).

Moreover, it can also be clearly observed from (3.20) that the voltage stability is increased under the stiffer grid condition (larger SCR) and/or with higher sending end voltage (larger V), as shown in Figure 3.4(a) and (b).

3.3.3   Impact of the local load

By considering the power consumption of the local load, it is known from Figure 3.2 that the active power transferred from Bus A to Bus B can be reduced as: $P = P_{gen} - P_{load}$. This reduction of the active power transfer would lead to a better voltage stability of the system. As an example given in Figure 3.4(b), with $P_{gen} = 1$ p.u. from the IBR, the active power delivered to the receiving end is reduced from 1 p.u. to 0.5 p.u. with 0.5 p.u. local load. It can be seen from the black solid line in Figure 3.4(b) that $\partial Q/\partial V \approx 0.8$ when $P = 0.5$ p.u. and $V = 0.93$ p.u., and the IBR can work stably in this scenario. Yet, $P = P_{gen} = 1$ p.u. is yielded when there is no local load, and $\partial Q/\partial V$ is almost reduced to zero with $P = 1$ p.u. and $V = 0.93$ p.u. (see the black solid line in Figure 3.4(b)). Evidently, the IBR can hardly be stabilized under this operating condition.

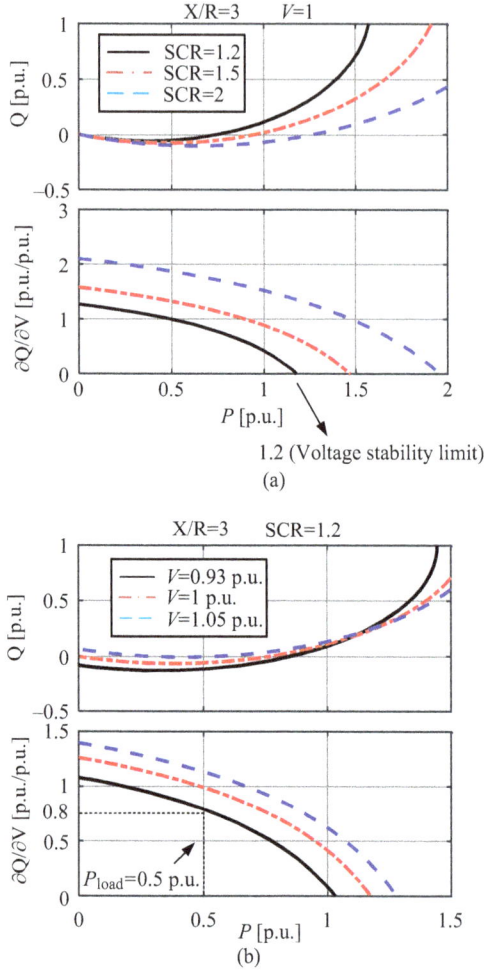

Figure 3.4   P–Q and P–∂Q/∂V *curves of the system with* X/R = 3, V_{th} = 1 p.u.
(a) V = 1 p.u., with different SCR. (b) SCR = 1.2, with different
sending end voltage V.

It is worth mentioning that the SCR of the power grid is not affected by local
loads at Bus A (see the definition of SCR from (3.13)). Yet, the previous analysis
has clearly demonstrated that the existence of local loads can increase the P_{max} of
the IBR. Therefore, SCR-based P_{max} assessment would lead to too pessimistic
result in the scenario with local loads. In a real power system, the load will be
distributed on different locations of the grid, and the assessment is not as simple as
in the illustration, but the general effect remains the same.

3.3.4   Impact of synchronous condenser

By considering a local SynCon, it is known from Figure 3.2 that the reactive power generated from the IBR can be expressed as: $Q_{gen} = Q_{SynCon} + Q$, which yields

$$\frac{\partial Q_{gen\_pu}}{\partial V_{pu}} = \frac{\partial Q_{SynCon\_pu}}{\partial V_{pu}} + \frac{\partial Q_{pu}}{\partial V_{pu}} \tag{3.22}$$

By following the similar procedure given by (3.16)–(3.20), $\partial Q_{SynCon\_pu}/\partial V_{pu}$ can be calculated as

$$\frac{\partial Q_{SynCon\_pu}}{\partial V_{pu}} = \frac{1}{X''_{d\_pu}} \left(2V_{pu} - \frac{V_{SynCon\_pu}}{\sqrt{1 - \sin^2\delta}} \right) \tag{3.23}$$

where X''_{d_pu} and V_{SynCon_pu} are the sub-transient reactance and the terminal voltage magnitude of the SynCon, respectively. δ represents the angle difference between voltages at Bus A and the terminal voltage of the SynCon. Since the SynCon does not generate nor absorb active power beyond the internal losses, $\delta \approx 0$ is yielded. Hence, (3.23) can be simplified as

$$\frac{\partial Q_{SynCon\_pu}}{\partial V_{pu}} = \frac{1}{X''_{d\_pu}} \left(2V_{pu} - V_{SynCon\_pu} \right) \tag{3.24}$$

Substituting (3.20) and (3.24) into (3.22), which yields

$$\begin{aligned}\frac{\partial Q_{gen\_pu}}{\partial V_{pu}} &= \frac{1}{X''_{d\_pu}} \left(2V_{pu} - V_{SynCon\_pu} \right) + \frac{1}{X_{th\_pu}} \left(2V_{pu} - \frac{V_{th\_pu}}{\sqrt{1 - \sin^2\theta}} \right) \\ &= SCR_{SynCon} \left(2V_{pu} - V_{SynCon\_pu} \right) + SCR_{th} \left(2V_{pu} - \frac{V_{th\_pu}}{\sqrt{1 - \sin^2\theta}} \right).\end{aligned} \tag{3.25}$$

where SCR_{SynCon} and SCR_{th} represent the SCR contributed by SynCon and the grid, respectively. Based on the definition given by (3.13), the SCR at the connection point of the IBR by considering the impact of SynCon can be calculated as

$$SCR = SCR_{SynCon} + SCR_{th}. \tag{3.26}$$

It is known from (3.25) that $\partial Q_{gen\_pu}/\partial V_{pu}$ is increased by a constant factor SCR_{SynCon} by adopting SynCon (assume $V_{SynCon} = V_{th} = 1$ p.u.), which is visualized by comparing the $\partial Q_{gen\_pu}/\partial V_{pu}$ curve with and without SynCon (red dashed and black solid line) in Figure 3.5. Therefore, the critical power angle where the voltage stability limit is reached is increased by adding the SynCon in the sending bus. Nevertheless, the angle stability limit defined in (3.10) cannot be exceeded regardless of the capacity of the added SynCon. Hence, the large transmission line impedance would still impose significant constraints on the P_{max} of the IBR, see (3.12).

Therefore, the improvement of the P_{max} of the IBR by adding local SynCon is usually not that significant compared with the case with the reduced transmission line impedance, even though they end up with the same SCR. As an example given

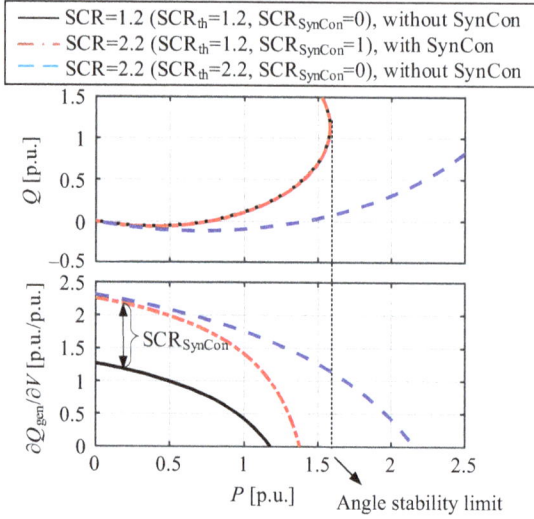

Figure 3.5   P–Q and P–∂Q/∂V curves of the system with or without SynCon. X/R = 3, V = V_{th} = V_{SynCon} = 1 p.u.

by red and blue lines in Figure 3.5, with the same SCR = 2.2, the power transfer capability of the IBR with smaller transmission line impedance but no local SynCon (SCR$_{th}$ = 2.2, SCR$_{SynCon}$ = 0) is higher than that with larger transmission line impedance and local SynCon (SCR$_{th}$ = 1.2, SCR$_{SynCon}$=1).

3.3.5   Discussion

Based on previous analysis, it turns out that inaccurate P_{max} prediction might be yielded if the assessment is performed solely based on the value of SCR, as SCR itself cannot fully reflect the impact of operating voltage level, local loads, and other controlled units (like SynCon) on the P_{max} of the IBR; as shown in Figures 3.4(b) and 3.5. In contrast, those impacts can be characterized by different values of ∂Q/∂V. Therefore, the minimum ∂Q/∂V, rather than minimum SCR, can be used as a metric for assessing P_{max} of the IBR with grid-following control. The value of minimum ∂Q/∂V can be calculated by the TSO based on max production and minimum load predictions.

3.4   Simulation results

Nonlinear time-domain simulations based on PSCAD/EMTDC are carried out to verify the ∂Q/∂V-based power transfer capability prediction in Section 3.3. It should be emphasized that ∂Q/∂V > 0 only indicates that the IBR is operated within the physical constraint (voltage stability limit) imposed by the ac system, careful controller parameters tunning of the IBR is still needed to guarantee its stable operation.

1. *Impact of local loads under same SCR*

   Figure 3.6 shows the simulation results of the IBR connecting to the grid with SCR = 1.2, $X/R = 3$, $V_{th} = 1$ p.u. and $V = 0.93$ p.u. It can be seen from Figure 3.6(a) that the IBR becomes unstable when generating 1 p.u. active power with the absence of local loads, due to the fact that $\partial Q/\partial V \approx 0$ is predicted in Figure 3.4(b). Yet, it is also known from Figure 3.4(b) that $\partial Q/\partial V$ is increased to 0.8 when there is 0.5 p.u. local loads, and the IBR can work stably in this scenario, which is verified by simulation results given in

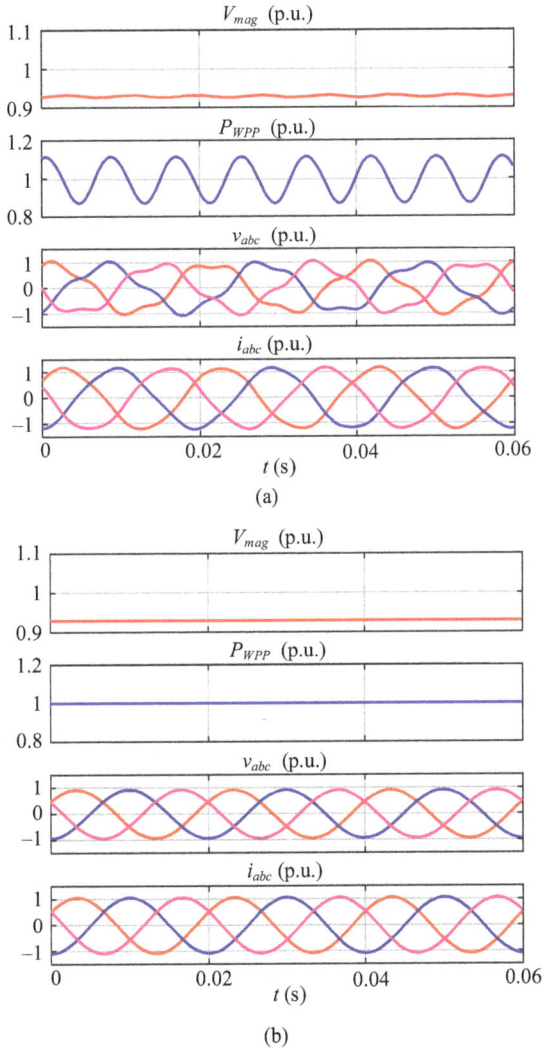

Figure 3.6 *Simulation results of the IBR connecting to the grid with SCR = 1.2, X/R = 3, V_{th} = 1 p.u. and V = 0.93 p.u. (a) Without local load, unstable. (b) With 0.5 p.u. local load, stable.*

Figure 3.6(b). The simulation results given in Figure 3.6 verify the power transfer capability predictions given in Figure 3.4(b) considering the impact of local loads under the same SCR.

2. Impact of SynCon under the same SCR

    Figure 3.7 shows the simulation results of the IBR connecting to the grid with SCR = 2.2. Two scenarios corresponding to Figure 3.5, i.e., with/without local

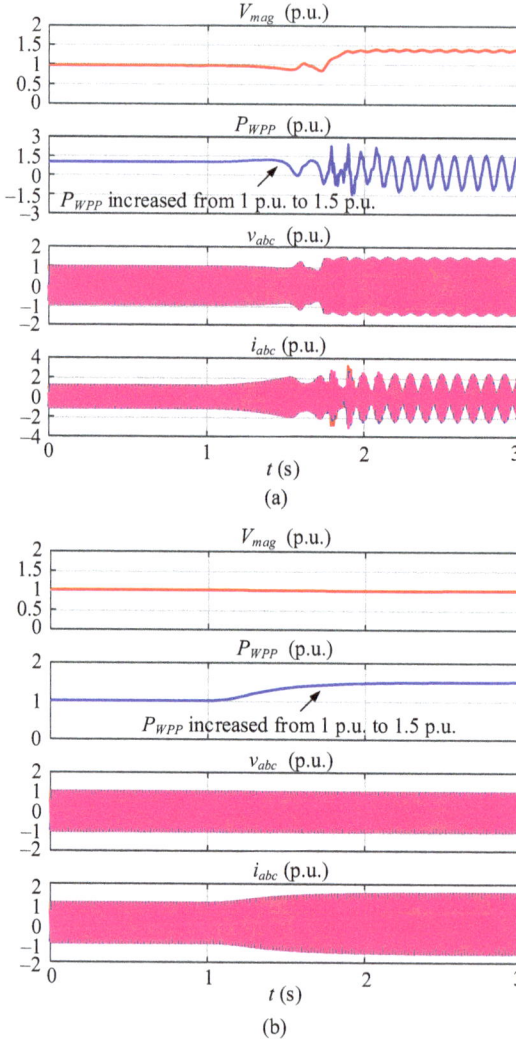

Figure 3.7   *Simulation results of the IBR connecting to the grid with SCR = 2.2, X/R = 3, V_{th} = 1 p.u. and V = 1 p.u. (a) With larger transmission line impedance and local SynCon (SCR_{th} = 1.2, SCR_{SynCon} = 1), unstable. (b) With a smaller transmission line impedance and no local SynCon (SCR_{th} = 2.2, SCR_{SynCon} = 0), stable.*

SynCon, are considered. It can be seen from Figure 3.7(a) that IBR becomes unstable when transferring 1.5 p.u. active power with larger transmission line impedance and local SynCon ($SCR_{th} = 1.2$, $SCR_{SynCon} = 1$), due to the negative $\partial Q/\partial V$ predicted by the red dash-dotted line in Figure 3.5. Yet, the blue dashed line in Figure 3.5 indicates $\partial Q/\partial V$ can be turned to positive at $P = 1.5$ p.u. if the transmission line impedance is reduced, even if local SynCon is not adopted ($SCR_{th} = 2.2$, $SCR_{SynCon} = 0$). The IBR can thus operate stably in this scenario, as demonstrated by simulation results given in Figure 3.7(b).

It is worth mentioning that the motivation for simulating IBR with $P = 1.5$ p.u. is merely to validate the power transfer capability predicted in Figure 3.5. In practice, the IBR normally does not have the 1.5 p.u. overload capability due to the hardware limit.

3.5 Discussion

While this chapter demonstrates the benefit of utilizing $\partial Q/\partial V$ (voltage sensitivity) based metric for stability screening, several challenges persist that are worth further investigation:

1. *Industrial implementation*: The short-circuit calculation has become a standard feature in many commercial software packages, offering a simplified means of calculating the SCR. Nevertheless, as far as the authors are aware, few commercial software packages support the calculation $\partial Q/\partial V$ (voltage sensitivity), thereby presenting challenges for industries, particularly TSOs that manage the large and complex electrical networks, to implement the $\partial Q/\partial V$-based metric.
2. *Threshold:* While there is a common acceptance of the boundary value of the SCR, typically ranging from 1.5 to 2, below which the system can be classified as a "weak grid," a similar threshold value has not been established for $\partial Q/\partial V$ that distinguishes between weak and stiff grids. Therefore, more investigation is needed in this direction.
3. *Applicability for a system with mixed grid-following/forming IBRs*: All the analyses carried out in the chapter consider a grid-following IBR that is modeled as a PQ source. However, for grid-forming IBR that is essentially a PV source, the power transfer capability is not constrained by $\partial Q/\partial V > 0$ (but still constrained by $\partial P/\partial \theta > 0$), as the grid-forming IBR can establish the voltage independently, rather than relying on the dedicated reactive power control. Therefore, determining the minimal value of $\partial Q/\partial V$ that can guarantee the stable operation of the system with mixed grid-following/forming IBRs would be more challenging and worth further investigation.

3.6 Conclusion

This chapter points out that the minimum $\partial Q/\partial V$ (voltage sensitivity) is a more appropriate metric for assessing the power transfer capability of the grid-following IBR compared to the minimum SCR. Simulation results are given to demonstrate the effectiveness of the proposed control method.

References

[1]   S. Huang, J. Schmall, J. Conto, J. Adams, Y. Zhang, and C. Carter, "Voltage control challenges on weak grids with high penetration of wind generation: ERCOT experience," in *Proceedings of the IEEE PES General Meeting*, 2012, pp. 1–7.

[2]   N. Modi, B. Badrzadeh, A. Halley, A. Louis, and A. Jalali, Operational Manifestation of Low System Strength Conditions – Australian Experience, Cigre Session, 2020, C2-124.

[3]   Cigre Working Group B4.62. TB 671: "Connection of wind farms to weak AC networks," Technical Report, Cigre, 2016.

[4]   P. Kundur, *Power System Stability and Control*. New York, NY: McGraw-Hill, 1994.

[5]   T. Cutsem and T. Vournas, *Voltage Stability of Electric Power Systems*. New York, NY: Springer, 2005.

[6]   NERC Reliability Guideline, "Integrating inverter-based resources into low short circuit strength systems," NERC, Atlanta, GA, December 2017. www.nerc.com.

[7]   NERC White Chapter, "Short-circuit modeling and system strength," NERC, Atlanta, GA, February 2018. www.nerc.com.

[8]   Australian Energy Market Operator Limited (AEMO), "System strength in the NEM explained", AEMO, Mar. 2020. https://aemo.com.au/-/media/files/electricity/nem/system-strength-explained.pdf

[9]   Y. Zhang and A. M. Gole, "Quantifying the contribution of dynamic reactive power compensators on system strength at LCC-HVdc converter terminals," *IEEE Transactions on Power Delivery*, early access, doi: 10.1109/TPWRD.2021.3063153.

[10]   J. Matevosyan, "Weak grid experiences in ERCOT," Presented in Western Electricity Coordinating Council (WECC) Workshop, October, 2021. https://www.esig.energy/download/wecc-workshop-weak-grid-experiences-in-ercot-julia-matevosyan/

[11]   J. Z. Zhou and A. M. Gole, "VSC transmission limitations imposed by AC system strength and AC impedance characteristics," in *10th International Conference on AC DC Power Transmission*, Birmingham, UK, December 2012.

[12]   S. Achilles, "Weak grid connection of IBR, why are we still talking about this?," Presented in G-PST/ESIG Webinar, November2021. https://www.esig.energy/event/g-pst-esig-webinar-series-weak-grid-connection-of-ibr-why-are-we-still-talking-about-this/

[13]   T. Lund, H. Wu, H. Soltani, J. G. Nielsen, G. K. Andersen and X. Wang, "Operating wind power plants under weak grid conditions considering voltage stability constraints," *IEEE Transactions on Power Electronics*, 2022, vol. 37, no. 12, pp. 15482–15492, 2022.

[14]   Y. W. Li and C.-N. Kao, "An accurate power control strategy for power electronics-interfaced distributed generation units operating in a low voltage multibus microgrid," *IEEE Transactions on Power Electronics*, vol. 24, no. 12, pp. 2977–2988, 2009.

[15]   T. Lund., "Analysis of Distribution Systems with a High Penetration of Distributed Generation," PhD Thesis. Ørsted, Technical University of Denmark, 2007.

Chapter 4

Dynamic model reduction of power networks for fast assessment of power system strength – part 1: classical techniques

Lahiru Aththanayake Adhikarige[1], Ameen Gargoom[1] and Nasser Hosseinzadeh[1]

Abstract

The system strength assessment (SSA) is essential for contemporary power systems integrated with various invertor-based generators (IBGs). These IBGs intrinsically interfaced by power electronic converters and their intermittent energy sources, such as wind and solar, create a new type of challenge to power system operators and researchers. Moreover, ever-increasing demands have overstretched the operating point (OP) of power systems, which is now operating closer to its instability point. Furthermore, the loads in the distribution system which were regarded as constant loads SSA can no longer be considered as such with the advent of active distribution networks (ADNs). Concerning these factors, a swift and precise SSA is needed to operate the power system accurately without the loss of power to the customer. The SSA studies are typically conducted on a specific area of the power system, especially in the weak parts of the system, which is called the "study system" (SS). The area outside the SS is called the "external system" (ES), which shall be represented as a reduced order model in the SSA for accelerated results. With the evolution of power systems, the model reduction techniques branched into two fields, i.e., classical techniques and measurement-based techniques. Both techniques have their merits and drawbacks. Therefore, these two broad categories of power system reduction techniques are discussed in two chapters. In this chapter, the reduction techniques for dynamic equivalent models of the ES from the classical techniques are discussed. A case study is performed on a simplified Australian 14 generator model to show the potential of classical techniques in the reduction of power systems and their strengths limitations and limitations in applying to modern IBRs rich power networks are examined.

[1]Centre for Smart Power and Energy Research, School of Engineering, Deakin University, Australia

4.1   Introduction to system strength

The transition from fossil fuel to renewable energy allowed room for the mass installation of wind power plants (WPPs) and solar power plants (SPPs). The behavior of the power systems consisting of these power plants is distinct from a conventional synchronous generator-driven power system. This is because the power electronic interface through which IBGs are connected to the grid determines the nature of the power system as their penetration dominates. Some of the most common problems associated with renewable-rich networks are harmonic distortion, lack of primary frequency response, reduced inertia, and insufficient voltage support [1]. The danger of an IBG-dominated power system was first experienced in the South Australian blackout in 2016 [2]. The approach to resolve the issues associated with increased IBG penetration can be looked at in two ways, i.e., proposing reforms to the network structure to cope with undesirable effects and quantifying the impact of IBG penetration on the stability of the system. The system strength evaluation is one such technique that is performed to determine the voltage stability of a particular bus in the system.

The system strength of a power system is defined as a characteristic that relates to the size of the voltage excursion in a particular bus after a disturbance [3]. The most common index that evaluates the system strength is called the short circuit ratio (SCR). The SCR for a renewable energy source (RES)-connected bus is defined as the ratio of the fault level to real power injection from the RES [4]. The equation for the SCR is given by,

$$SCR = \frac{S_{ac,i}}{P_{d,i}} \tag{4.1}$$

where $s_{ac,i}$ is the fault level (or the short circuit capacity) at the ith bus and $P_{d,i}$ is the rated power capacity of the RES connected at bus i.

The expression for the SCR is closely related to the expression of the voltage stability of a particular bus resulting from the singularity of the Jacobian matrix [5]. This ratio for the voltage stability can be written as,

$$\frac{|S_{a,c,i}|}{|S_i^*|} = \frac{|V_i|^2}{|S_i^*||Z_i|} = 1 \tag{4.2}$$

where S_i^* is the complex power injected at bus i, $S_{a,c,i}$ is the fault level at bus i, V_i is the voltage magnitude at RES connected bus, and Z_i is the impedance between buses s and i.

From (4.2), bus i is voltage stable if the ratio is greater than 1, it is marginally stable if the ratio is equal to one and bus i is voltage unstable if the ratio is less than one. By comparison of (4.1) and (4.2), a similarity can be observed in the complex power injection at bus i and the rated power capacity of the RES. Therefore, a RES-connected bus must have an SCR ratio of above 1 for it to be voltage stable.

However, there is one major problem in the assessment of the system strength in large power systems regarding their dimensionality. Since the system strength is evaluated by the SCR, the fault-level calculation is a tedious task in geographically expanded networks. Calculation of the short circuit capacity or the fault level necessitates the calculation of symmetrical short circuit current in the point of interest. This is given by the following equation:

$$I_{SC} = \frac{V_i}{\sqrt{3}Z_{eq}} \tag{4.3}$$

where I_{SC} is the short circuit current, V_i is the nominal voltage at bus i, and Z_{eq} is the equivalent impedance in the fault loop.

The value of Z_{eq} must include all the impedances in the fault loop starting from the points where sources are connected to the place where the fault is grounded. Then the equivalent impedance is calculated as seen from the short-circuited point towards the upstream network. In places where upstream impedances are unknown, it is reasonable to assume the source voltages are 80% of V_i [6].

Equation (4.3) suggests that for large-interconnected networks, there must be a reduction strategy to equivalence the ES impedance so that it is made simpler to obtain the short circuit current. Therefore, in the upcoming parts of this chapter, model reduction strategies for the ES of a power system considering both conventional power systems dominated by synchronous generators (SGs) as well as contemporary power systems integrated with IBGs are comprehensively discussed.

4.2   Model reduction strategies

4.2.1   Background

Power system SSAs are typically focused on a specific area of the power system known as the study system (SS). This is because the interest of SSA studies is rarely focused on the entire power systems, but rather focused only on the weak areas of the power system. However, due to the interconnected nature of large power systems, it is computationally challenging to estimate the system strength considering the entire power system. Therefore, model reduction strategies have been introduced to reduce the system attached to the SS, which is known as the external system (ES) and represent it as a reduced order model. In this regard, the model reduction strategies are broadly categorized into two, i.e., classical model reduction techniques and measurement-based model reduction techniques. The main difference between these two is the model dependency in the former and the application of measurement data for obtaining the reduced order model in the latter. The separation of the power system into the SS and the ES for the SSA is depicted in Figure 4.1.

According to Figure 4.1, the SS and the ES will be interconnected by tie-lines at their respective boundary buses, which inherit properties such as voltage (V), voltage angle (δ), and frequency (f). Active (P) and reactive power (Q) flows will be present during the steady-state conditions as well as the dynamic conditions of the power system. Both the SS and the ES consist of load buses and generator buses that include a mixture of generation resources based on SGs or IBGs such as wind power plants (WPPs) and (SPPs).

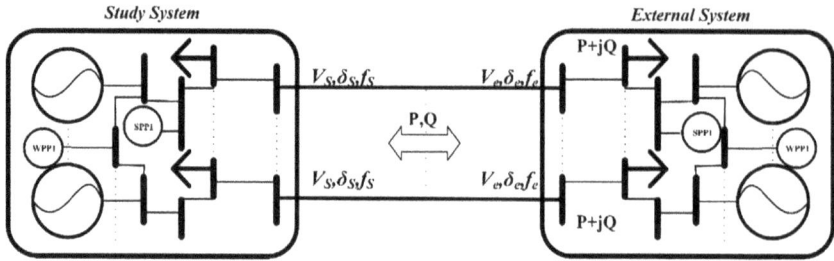

Figure 4.1   Block diagram representation of the SS and the ES

4.2.2   Overview

The early stages of the power systems were primarily dominated by the SGs. The penetration levels of IBGs in such systems were insignificant in determining the overall system dynamics. Therefore, the IBGs were often regarded as negative loads in the model reduction techniques at the early stages. This is known as the generator netting process in PSSE software [7]. However, as the penetration of the IBGs increased, their dynamics were reflected in the overall dynamics of the power system. Therefore, disregarding their presence in the model reduction techniques led to inaccurate system strength results of the SS.

In SGs-dominated power systems, the model reduction strategies were relatively simple. The SS boundary buses are identified with their respective tie lines that interconnect them to the ES. Then a topological model reduction is performed on the ES model, hence known as the classical model reduction techniques. However, as the penetration of the IBGs increased, together with the expansion of the power systems, three challenges were raised to the model reduction techniques. The first is the presence of the IBGs, which are typically based on intermittent renewable energy sources such as wind and solar that fluctuate the power system operating point frequently. As a result, tedious model reduction processes involved in classical techniques needed to be repeated to account for those operating point changes. The second is that competitive energy markets among the utility companies resulted in the restriction of model parameters of the IBGs [8]. Therefore, the option of a model-dependent reduction strategy was unavailable. The third is the increased dimensionality of the power system over the years that evoked complex nonlinear equations, extensive interconnections, geographical expansions, bi-directional power flows restricted the application of tedious classical techniques due to memory and computational power constraints [9]. As a result of all these three factors, plus the extensive installation of phasor measurement units (PMUs) [10] in the power system, the model reduction techniques slowly shifted to measurement-based strategies. However, the classical techniques are still in use in many power system simulations' software like PSSE and Digsilent, as well as they form a strong foundation for data-driven techniques, which emphasizes their soundness in the model reduction techniques. Accordingly, the discussion of model reduction strategies is divided into two, where Part 1 discusses the classical reduction strategies and Part 2 discusses the data-driven model reduction strategies.

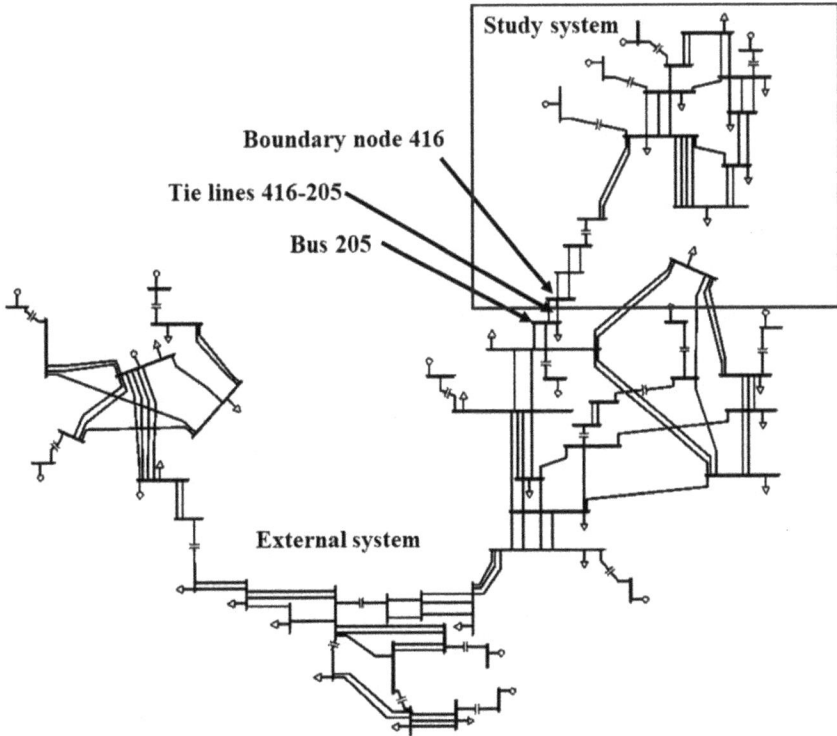

Figure 4.2   External and study systems representation of AU14G model

A representation of the ES and the SS for South Australian network in Simplified Australia 14 Generator System (AU14G) is given in Figure 4.2 [11].

AU14G is a simplified version of eastern Australian power system (NEM) in 2014. Figure 4.2 shows the electrical diagram of 5 eastern and southern states of Australia which consist of 14 generators interconnected with tie-lines. The area shown inside the box represents the Queensland (QLD) system, which in this case is considered as the SS. By selecting the QLD system as the SS, the area outside the box eventually becomes the ES. The two tie-lines interconnect the SS and the ES terminating at their respective boundary buses.

4.2.3   Classical reduction techniques

The classical reduction techniques can be primarily divided into two as static and dynamic. The static equivalent techniques focus on the steady state power flow between the SS and the ES. Hence, they can be used for power system planning and static security assessment. The dynamic techniques can be used for three applications, i.e., planning studies, dynamic security assessment, and real-time simulations. The key feature of classical techniques is that they explicitly focus on the model of the power system. The power system is a complex nonlinear structure that can be represented by a set of first-order nonlinear ordinary differential equations and algebraic equations as follows.

$$\dot{x} = f(x, u, t) \tag{4.4}$$

$$y = g(x, u) \tag{4.5}$$

where x is a set of state variables, u is the vector of inputs, y is the vector of outputs, and t is the time.

All classical techniques rely upon an extended set of equations described generally by (4.4) and (4.5), such as generator dynamic equations and load flow equations.

4.2.3.1   Classical static equivalent techniques

Classical techniques were introduced to determine the power system stability before major expansions in certain areas of the power system. Classical static equivalent techniques were first introduced in the 1930s, when Kron developed a method to retain the power balance in buses with current injections and reduce the rest of the network using tensor analysis [12]. This method suggests that the effect of a reduced node in a tensor can be reflected as impedance corrections in all other nodes. The techniques which were later developed such as the Ward technique and radial equivalent independent (REI) technique were derived from Kron's technique, which is fundamentally based on Gaussian elimination [13].

The Ward technique was able to even further reduce the external system to boundary nodes [14]. The fundamental idea behind the Ward technique is to represent the eliminated generators and loads as equivalent generators or loads connected at the terminal buses. This transformation was performed by calculating a set of power transfer distribution factors at the boundary nodes.

For any boundary node i, the equivalent powers P^{EQ} and Q^{EQ} resulting from all branch connections ij can be calculated by (4.6) and (4.7)

$$P_i^{EQ} = \sum_{j=1}^{n} \left[\left(V_i^0 \right)^2 \left(g_{ij} + g_{i0} \right) - V_i^0 V_j^0 \left(g_{ij} \times \cos \theta_{ij} + b_{ij} \times \sin \theta_{ij} \right) \right] \tag{4.6}$$

$$Q_i^{EQ} = \sum_{j=1}^{n} \left[-\left(V_i^0 \right)^2 \left(b_{ij} + b_{i0} \right) + V_i^0 V_j^0 \left(b_{ij} \times \cos \theta_{ij} - g_{ij} \times \sin \theta_{ij} \right) \right] \tag{4.7}$$

However, one limitation in the original Ward's technique is that it was unable to accurately determine reactive power injections at the boundary nodes for disturbances in the SS. Therefore, the extended Ward technique was introduced to address this issue [15].

Thereafter, to guarantee a lossless equivalent representation of the ES, the REI technique was introduced by Dimo [16]. The first step of the REI reduction is to disconnect the loads from the zero potential nodes and connect it to a fictitious node. To balance the change in current flows, a new equivalent admittance $-Y_{eo'}$ is connected to the fictitious node calculated by (4.8):

$$\frac{I_e^2}{Y_{eo'}} = \sum_s \frac{I_s^2}{Y_{so'}} \tag{4.8}$$

where

$$Y_{so'} = \frac{-S_s^*}{|V_s|^2} \tag{4.9}$$

$$I_s = \frac{-S_s^*}{V_s} \tag{4.10}$$

$$I_e = \sum_{i=1}^{s} I_i \tag{4.11}$$

Hence,

$$V_{o'} = I_e \times Y_{eo'} \tag{4.12}$$

In this technique, a zero-power balance network is assured by selecting a set of fictitious nodes, where the buses to be eliminated are initially connected to [17]. Later, in the Gaussian elimination process, all redundant buses and fictitious nodes are eliminated.

The Gaussian elimination process can be expressed in the following way. Let us take one load node S in the system. This node consists of one branch connected to fictitious node O' as well as branch/branches connected to other nodes. The Gauss elimination technique suggests that the elimination of a load node S reflects additional branches in the remaining nodes connected to this load node S. Therefore, the updated admittances of the remaining nodes m and n connected to node S can be written as,

$$Y_{mn,new} = Y_{mn,former} - \frac{Y_{sm}Y_{ms}}{Y_{ss}} \tag{4.13}$$

where Y_{ss} represents sum of all radial admittances connected to node S.

The transformation given by (4.13) is equivalent to a star-delta transformation between nodes m, n, and S.

The works of these three techniques established the foundation for most model reduction techniques existing today, including commercial software packages like PSSE [7]. It should be noted that the Gaussian elimination should only be applied to redundant buses and non-border buses. This is because the total current injection to the system must be preserved in the reduced system while preserving the network structure.

4.2.3.2   REI reduction in IEEE 14 bus system

In this section, the REI technique is applied to eliminate three redundant buses (buses 10, 11, and 14) and two generator buses (buses 12 and 13) in the ES of the

IEEE 14 bus system. The IEEE 14 bus system is a standard test system that contains 14 buses, 11 loads, 7 machines, and 19 branches. As seen in Figure 4.4, the SS encloses buses 1–5 and the ES contains buses 5–14 in this test system. The SS contains 2 boundary buses (buses 4 and 5) interconnecting the three boundary buses

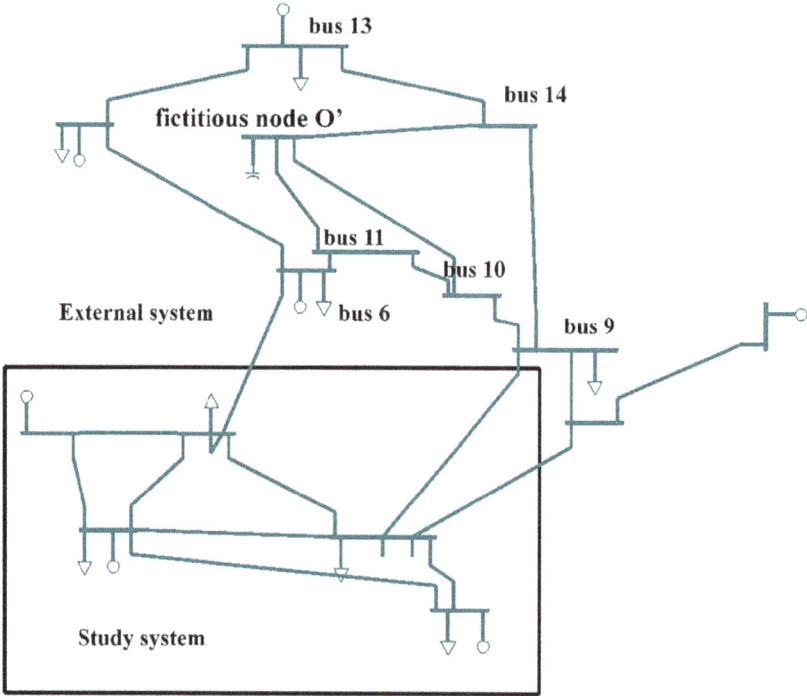

Figure 4.3   Representation of the SS and the ES in the unreduced

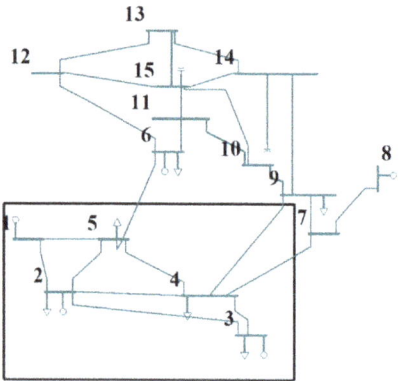

Figure 4.4   Representation of the fictitious REI node

Table 4.1   Load flow data and calculated shunt values of the buses to be eliminated

Bus no.	Voltage mag. (pu)	y_P (pu)	y_L(pu)	Total y (pu)	I_s
10	0.9141	0	$-0.1077 + j0.06941$	$-0.1077 + j0.06941$	$-0.0984 + j0.0634$
11	0.8546	0	$-0.7325 + j0.71$	$-0.7325 + j0.71$	$- 0.6260 + j0.6068$
12	1	$0.2 - j0.024$	$-0.135 + j0.058$	$0.0650 + j0.0340$	$0.0650 + j\ 0.0340$
13	1	$0.3 - j0.314$	$-0.061 + j0.016$	$0.2390 - j0.2980$	$0.2390 - j0.2980$
14	0.9657	0	$-0.21445 + j0.07506$	$-0.2145 - j0.0751$	$-0.2071 - j0.004$

of the ES (buses 6, 7, and 9) via three tie lines (5 and 6, 4–7, and 4–9). The ES consists of 4 generator buses (6, 8, 12, and 13) and 5 load buses (7, 9, 10, 11, and 14). Assuming that the system loads can be represented as constant impedances and generator power injections at buses 12 and 13 can be represented as negative loads, the REI reduction of the ES by retaining 3 boundary buses (buses 6, 7, and 9) and one generator bus (bus 8) of the ES is explained in the following steps. Before applying the REI technique, a load flow analysis is carried out on the unreduced system in Figure 4.3 and the load flow data respective to the ES buses are shown in Table 4.1.

Step 1: **Converting loads connected at the buses to be eliminated to shunt admittances.**

The first step in the REI technique is to perform a power flow solution and find the shunt admittances respective to each load connected at the redundant buses. In this step, power injected at buses 12 and 13 are regarded as negative loads. The results for the calculated shunts and their respective currents are summarized in Table 4.1.

Step 2: **Calculation of the REI quantities**

The REI quantities are calculated from (4.8) to (4.12) using the data given in Table 4.1:

$$I_{eq} = -0.6275 + j0.47807$$

$$\frac{I_{eq}^2}{Y_{e0}} = -0.5279 + j0.38202$$

$$\frac{(-0.6275 + j0.47807)^2}{Y_{e0}} = -0.5279 + j0.38202$$

$$Y_{e0} = 1.22126 - j0.02212$$

$$V_{e0} = 0.9636 < -38.34\text{deg}$$

From the calculated REI quantities given above, the loads at buses 10–14 are transferred to the fictitious bus 15 as shown in Figure 4.4.

As seen from Figure 4.4, the loads and generation at buses 10–14 are represented as an equivalent admittance connected at the fictitious node with a corresponding voltage and an angle. The lines connecting these buses to bus 15 have admittance values calculated in Table 4.1 (**Total y (pu)**).

Step 3: **Gauss elimination of redundant nodes.**

In this step, all redundant buses 10–15 are eliminated from the Gaussian elimination process given by (4.13). After step 2, the only component connected to these redundant buses other than the equivalent shunt connected to bus 15 is the switched shunt connected to bus 14. Before executing the Gaussian elimination process, this switched shunt is also converted to a fixed shunt from the respective load flow value injected by that to the bus. Therefore, the values given in Tables 4.2 and 4.3 are entered to construct the admittance matrix and reduce its size from the Gaussian elimination.

Table 4.2   Bus data for the ES (net conductance and net susceptance)

Bus no.	Net G	Net B
6	−0.1583	−0.1741
7	0	0
8	0	−0.234
9	−0.3281	−0.0056
10	0	0
11	0	0
12	0	0
13	0	0
14	0	0.1502
15	−1.2213	0.0221

Table 4.3   Line data for the ES (resistance, reactance, and line charging)

Bus no. from	Bus no. to	Line r	Line x	Line B
6	11	0.09498	0.1989	0
6	12	0.12291	0.25581	0
6	13	0.06615	0.13027	0
7	8	0	0.17615	0
7	9	0	0.11001	0
9	10	0.03181	0.0845	0
9	14	0.12711	0.27038	0
10	11	0.08205	0.19207	0
10	15	6.56081	4.22767	0
11	15	0.7046	0.68221	0
12	13	0.22092	0.19988	0
12	15	−7.17049	−0.41269	0
13	14	0.17093	0.34802	0
13	15	−1.77876	−2.75978	0
14	15	4.15417	1.454	0

The Gauss elimination was repeatedly executed to the admittance matrix of the ES constructed using data given in Tables 4.1 and 4.2. Ultimately, only buses 6, 7, 8, and 9 of the ES are retained in the admittance matrix. The resultant reduced admittance matrix (Y_{red}) after the Gaussian elimination process is given in Table 4.4.

As seen from Table 4.4, only 4 buses are retained in the reduced system, which corresponds to buses 6–9. The next step is to find actual branch values and bus values from Y_{red}.

Step 3: **Calculation of the branch values and the bus values of the reduced system.**

In this step, effective loads, branch admittances, and fixed shunts will be calculated for the reduced system. The branch values can easily be calculated by (4.14) from the elements of (Y_{red}), as shown in Table 4.5:

$$y_{ij} = -Y_{ij} = -Y_{ji} \tag{4.14}$$

Load/fixed shunt values are calculated by applying (4.15) to the elements of the reduced admittance matrix Y:

$$y_{io} = Y_{ii} - \sum_{k=1}^{n} y_{ik}; k \neq i \tag{4.15}$$

However, the shunt values given in Table 4.6 are not the final values to be connected to buses 6–9 since they already contain pre-existing components. These shunts are derived as the total current paths to zero potential at buses 6–9. To calculate additional shunt values needed to be added to each bus to reflect the reduced admittance matrix, the pre-existing admittances must be reduced from these results.

Fixed shunt values to be added to buses 6 and 9 considering the pre-existing components to reflect the reduced admittance matrix:

$$y_{6o} = 2.6979 - 34.2263i \text{MVA}$$

$$y_{9o} = 13.8683 - 77.83i \text{MV}$$

Step 4: **Power flow mismatch correction.**

After completing steps 1–4, if there is a mismatch in the power flows of the tie-lines compared to the unreduced system, it should be corrected by adding shunt

Table 4.4 Reduced admittance matrix

1.11069509 −3.12188955j	0+0i	0+0i	−1.242026 +2.60548686j
0+0i	0 −14.76706257j	0. +5.67697985j	0. +9.09008272j
0+0i	0. +5.67697985j	0. −5.91097985j	0+0
−1.242026 +2.60548686j	0. +9.09008272j	0+0i	1.05259852 −12.47382845j

Table 4.5   Reduced system branch values

From bus	To bus	R	X	B
6	9	0.14908	0.31274	0
7	8	0	0.17615	0
7	9	0	0.11001	0

Table 4.6   Values calculated for y_{io} from the reduced admittance matrix

Bus no.	y_{io} (MVA)
6	$-13.13309121 -j51.64026902$
7	0
8	0
9	$-18.94274858 - j77.82588615$

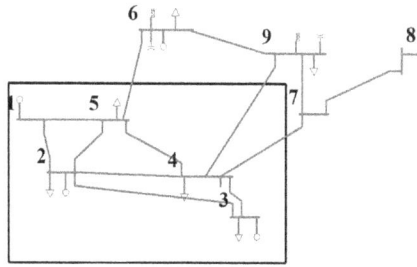

Figure 4.5   REI reduced equivalent system

components at respective boundary buses of the ES. In this reduction, power flow mismatches were observed in tie-lines 5 and 6 and 4–9. Therefore, the following additional shunt components were added at buses 6 and 9 of the ES:

$y_{6o} = 16.16 + 21.02i$MVA
$y_{9o} = 30.15 + 51.41i$MVA

The single-line diagram in Figure 4.5 shows the final REI-reduced equivalent system of the unreduced network.

Active power flows
The following plots provide the power flows in tie lines of the reduced system compared with the power flows in the original system.

This graph represents the active power flows in tie-lines 5 and 6 of the REI reduced system and the original system.

This graph represents the reactive power flows in tie-lines 5 and 6 of the REI reduced system and the original system. From Figures 4.6 and 4.7, the power flows in the reduced system are not precisely equal to the unreduced system. In the active

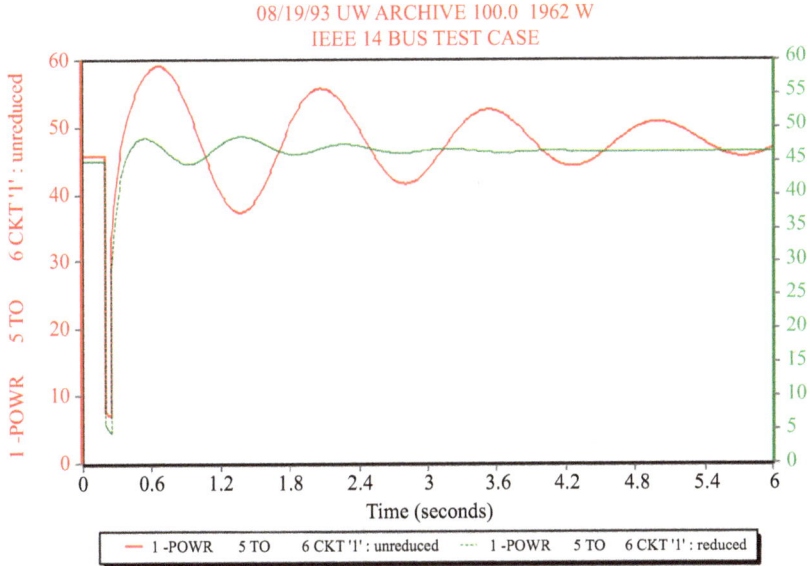

Figure 4.6 Active power flow in tie-lines 5 and 6 for a similar fault in the unreduced system and the reduced systems

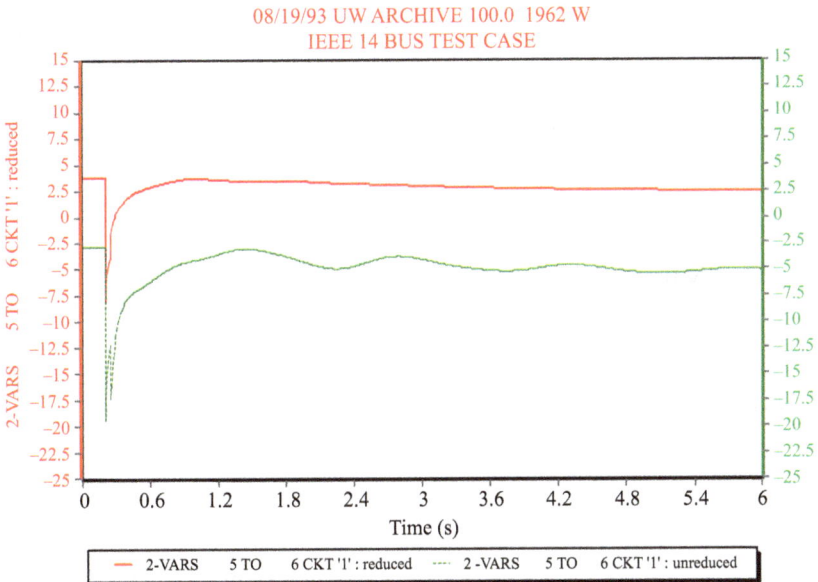

Figure 4.7 Reactive power flow in tie-lines 5 and 6 for a similar fault in the unreduced system and the reduced systems

power flow (Figure 4.6), there is a small steady-state deviation as well as large deviations in the post fault oscillations. The steady-state error is negligible whereas the deviations in oscillations could be a result of eliminating the dynamics of two generators in the ES. In the reactive power flow (Figure 4.7), there is approximately a 5 MVAR steady-state deviation as well as transient and post fault oscillations deviations. In this case, the error in the reactive power flow cannot be neglected for stability studies.

Although the Ward and the REI techniques are both derived from the Kron technique, a key difference between these two techniques is that the Ward technique considers generator buses as well in the reduction process. In contrast to the Ward technique, the REI technique adds one new bus (fictitious bus) to the network each time a set of load buses are eliminated. However, the output of these three techniques can only produce a static equivalent model. The application of the static equivalent techniques was limited as the power systems outgrew with increased complexity which necessitated dynamic stability analysis.

4.2.4   Classical dynamic equivalent techniques

Classical dynamic equivalent techniques are based on the structural details of the ES while they can perform dynamic stability analysis simulations such as transient stability analysis. The transient stability of the power system, also known as the large-disturbance rotor angle stability, determines the ability of the power system to regain its stability after a severe disturbance such as transmission line faults, tripping of a generator or tripping of a large load [18]. In the situation of a large disturbance in the SS, power flows through interconnected transmission lines between the two systems have an impact on the stability of the SS. Correspondingly, the reduced dynamic equivalent of the ES can speed up the convergence of the transient stability analysis simulation. In this regard, there are two main types of classical dynamic equivalent techniques, i.e., coherency-based techniques and power system model-based techniques.

4.2.4.1   Coherency-based techniques

The coherency-based technique is arguably the most effective classical dynamic equivalent technique because of its ability to preserve the oscillation modes of the original system in the reduced equivalent. Coherency is the ability of synchronous generators to swing together as a group of machines with constant angular deviations over a period of time [19]. There are three first steps to a coherency-based dynamic model reduction technique, i.e., identification of coherent generators, generator bus aggregation, and equivalent generator parameter identification.

4.2.4.1.1   Identification of coherent generators
The coherency of generators in a conventional power system can be determined in two ways, i.e., by exciting the power system with a disturbance and by calculating the angular deviations of synchronous generators or by singular perturbation technique [20]. The former is also called the signal-based method while the latter is called the slow-coherency technique.

Signal-based coherency identification

In the signal-based method, a clustering algorithm must be used to separate the coherent groups from the calculated angular deviation signals. The initial efforts of the signal-based method date to the 1970s, when a project conducted by the Electrical Power Research Institute (EPRI) published a paper that examines the identification of coherent generators using a linear network model and a classical generator representation [21]. The linear simulation algorithm explained in this paper uses a step increase in either ΔP_m or ΔP_l to simulate a disturbance. After that, it uses the trapezoidal integration rule to calculate bus voltage angles $\Delta\theta(t)$ and generator rotor angle deviations $\Delta\delta(t)$. Then, rotor speed deviations are calculated using $\Delta\dot{\delta}(t)$. Afterwards, a clustering algorithm is applied to identify the coherent generators from calculated generator internal voltage swing curves or generator terminal voltage swing curves. However, in this step, the calculation of machine angles was later replaced by measurement devices installed in different locations of the power systems.

Slow-coherency method

In the slow-coherency technique, a singular perturbation parameter, also known as a weak connection parameter, is used to separate the system swing modes into fast and slow time scales. The fast time scales represent the oscillations of machines within a coherent group while the slow time scales represent the oscillations between coherent groups. Let us take a system with n synchronous machines represented by classical models and N buses. Then the equations for the synchronous generator dynamics can be written as,

$$M\ddot{\delta} = f(\delta, V) \tag{4.16}$$

$$\dot{\delta} = \Omega\omega \tag{4.17}$$

Equations for power balance of network algebraic equations can be written as follows:

$$0 = g(\delta, V) \tag{4.18}$$

where $\Omega = 2\pi f$, δ is the machine internal rotor angle, M is the machine inertia, V is the machine terminal bus voltage/load bus voltage, f represents the machine swing equation, and g represents the load flow equation. By linearizing (4.16)–(4.18) around an operating point, (4.19) is obtained:

$$\begin{bmatrix} M\Delta\ddot{\delta} \\ 0 \end{bmatrix} = \begin{bmatrix} J_A & J_B \\ J_C & J_D \end{bmatrix} \begin{bmatrix} \Delta\delta \\ \Delta V \end{bmatrix} \tag{4.19}$$

where, $J_A = \frac{\partial f(\delta,V)}{\Delta\delta}$, $J_B = \frac{\partial f(\delta,V)}{\Delta V}$, $J_C = \frac{\partial g(\delta,V)}{\Delta\delta}$, $J_D = \frac{\partial g(\delta,V)}{\Delta V}$ and Δ represents a step change in the operating point from δ_0 and V_0. In (4.16), machine damping factor (D) and input mechanical power deviation ΔP_m are neglected. It is also assumed that the linearization was done around an equilibrium point which results $\Delta\omega = 0$.

Thus, (4.19) can be simplified into (4.20) and (4.21):

$$M\Delta\ddot{\delta} = J_A\Delta\delta + J_B\Delta V \tag{4.20}$$

$$0 = J_C\Delta\delta + J_D\Delta V \tag{4.21}$$

Details about matrices J_A, J_B, J_C, and J_D are skipped for brevity and can be found in [19].

The Kron reduced model of the system given by (4.20) and (4.21) can be written by (4.22):

$$M\Delta\ddot{\delta} = A\Delta\delta \tag{4.22}$$

where

$$A = J_A - J_B J_D^{-1} J_C \tag{4.23}$$

From (4.22), it can be noticed that the dynamics of the system is defined by matrix $M^{-1}A$, which is called the synchronizing coefficient matrix. Matrix A is also a square matrix $(n \times n)$ with elements representing the synchronizing torques (A_{ij}) between generator i and j in the system. Its diagonal elements A_{ii} represent the negative square sum of elements in row i. Construction of the matrix A from the reduced admittance matrix Y_{red} is explained below.

After the elimination of the redundant buses in the system by (4.22), matrix equations of the power system will take the form given by (4.24):

$$
\begin{bmatrix} I_1 \\ I_2 \\ \cdot \\ \cdot \\ I_n \end{bmatrix} = \begin{bmatrix} Y_{11} & Y_{rg} & \cdots & Y_{1n} \\ Y_{21} & Y_{22} & \cdots & Y_{2n} \\ & & \cdot & \\ Y_{n1} & Y_{n2} & \cdots & Y_{nn} \end{bmatrix} \begin{bmatrix} E_1 \\ E_2 \\ \cdot \\ \cdot \\ E_n \end{bmatrix} \tag{4.24}
$$

where E is the internal voltages of machines, I is the current injection from machines to the network, n is the number of machines, and Y_{nn} is the reduced admittances connecting internal nodes of machines n and n. The equation for complex power generated by any machine k can be written as,

$$S_k = P_k + jQ_k = E_k \times I_k^* \tag{4.25}$$

The current injected by any generator k can be expressed as a multiplication of the internal voltages of generators connected to it and the admittances between them as follows:

$$I_k = \sum_{i=1}^{n} Y_{ki} E_i \tag{4.26}$$

where

$$Y_{ki} = G_{ki} + jB_{ki} \tag{4.27}$$

By substituting (4.26), (4.25) can be updated as follows:

$$S_k = E_k \sum_{i=1}^{n} Y_{ki}^* E_i^*$$ (4.28)

In (4.28), real and reactive power injections from generators can be separately written as

$$P_k = E_k \sum_{i=1}^{n} E_i[B_{ki}\sin(\delta_k - \delta_i) + G_{ki}\cos(\delta_k - \delta_i)]$$ (4.29)

$$Q_k = E_k \sum_{i=1}^{n} E_i[G_{ki}\sin(\delta_k - \delta_i) - B_{ki}\cos(\delta_k - \delta_i)]$$ (4.30)

The matrix A given in (4.22) is constructed using the expressions given by (4.29) and (4.30).

The system oscillations are closely related to square roots of the eigenvalues of $M^{-1}A$ and the generator participation factors are defined by eigenvectors of $M^{-1}A$ [22].

The effect of slow coherency is due to time-scale separation of the system. One reason for this is the high impedance connections in long transmission lines connecting different areas and low impedance connections in transmission lines within the same area. Moreover, coherent areas exhibit a higher number of connections compared to the number of connections within different coherent areas. By introducing this ε to the expression of admittance matrix J_D, (4.23) transforms into (4.31):

$$A = J_A - J_B\left(J_D^I + \varepsilon J_D^E\right)^{-1} J_C$$ (4.31)

$$A = J_A - J_B\left(J_D^I\right)^{-1} J_C - J_B\left(\varepsilon J_D^E\right)^{-1} J_C$$ (4.32)

From (4.32), the expression for A can be written in the general form as

$$A = A^I + \varepsilon A^E$$ (4.33)

where $A^I = J_A - J_B\left(J_D^I\right)^{-1} J_C$ and $A^E = J_B\left(J_D^E\right)^{-1} J_C$.

Thus, the two-time scale representation of the system given by (4.22) can be expanded as follows:

$$M\Delta\ddot{\delta} = \left(A^I + \varepsilon A^E\right)\Delta\delta$$ (4.34)

Where ε is called the singular perturbation parameter or weak connection parameter, A^I represents the matrix of internal connections within a coherent area and A^E represents the matrix of external connections within different coherent areas. To illustrate the time variables in (4.34) in the explicit form, slow variables and fast variables are transformed using (4.35) and (4.36):

$$y_\alpha = \frac{\sum_{i=1}^{n_a} m_i \Delta \delta_i}{\sum_{i=1}^{\alpha} m_i} \quad \alpha = 1, 2, \ldots \ldots, r \tag{4.35}$$

where n_α is the number of machines in the area α, m_i is the inertia of the machine i, and y_α is called inertia weighted aggregate variable.

The fast variable z_k^α denoting oscillations within the same area can be expressed as

$$z_{k-1}^\alpha = \Delta \delta_k - \Delta \delta_j k = 2 \ldots \ldots . n_\alpha \tag{4.36}$$

where $\Delta \delta_j$ is the rotor angle deviation of the reference machine in area α which has n_α number of machines and $\Delta \delta_k$ is the machine rotor angle deviation of the machine in area α considered for clustering. Thus, the two time-scale separated forms of generator rotor angle deviations can be explicitly written by (4.37):

$$\begin{bmatrix} y \\ z \end{bmatrix} = \begin{bmatrix} C \\ G \end{bmatrix} \Delta \delta \tag{4.37}$$

Details of matrices C and G are given in [19].

After integrating variables y and z into the Kron reduced model given by (4.37), the system equations are transformed into (4.38) and (4.39)

$$M_a \ddot{y} = \varepsilon A_a y + \varepsilon A_{ad} z \tag{4.38}$$

$$M_d \ddot{z} = \varepsilon A_{da} y + (A_d + \varepsilon A_{dd}) z \tag{4.39}$$

Details of the matrices M_a, M_d, A_a, A_{ad}, A_{da}, A_d, and A_{dd} can be found in [19].

The system given by (4.38) and (4.39) is called the standard singular perturbed form. If the fast variables are set to zero in the standard singular perturbed form, then the remaining network is called the inertial aggregated model. It is the zeroth-order approximation of the singular perturbed form, where internal nodes of generators in a coherent group are interconnected with infinite admittances. Hence, the coherent generators tend to excite together, in the same phase. Meanwhile, the slow-coherency technique is the impedance-corrected model of the inertial aggregated model since coherent generators are no longer interconnected with infinite admittances [23]. This is also called the first-order representation of the singular perturbed form of the system. The other branch of classical dynamic equivalencing can be linked to dimensionality reductions in linear control theory.

Slow-coherency method in wind integrated power system models
The analysis given by (4.21)–(4.53) has been expanded to power systems consisting of Doubly Fed Induction Generators (DFIG) in [24,25]. In these two papers, a linearized model for the wind-integrated power system is obtained before applying the two-time scale analysis. Subsequently, the linearized dynamic equation of a DFIG-based wind power plant model can be written by (4.40) [24]:

$$\Delta \dot{Z} = A \Delta Z + B_1 |\Delta V_n| \tag{4.40}$$

where V_n is the voltage of the bus where the WPP is connected and Z is the state vector of the DFIG WPP. Details of the matrices A and B_1 can be found in [24].

Moreover, the linearized power outputs of the DFIG WPP can be written as

$$
\begin{bmatrix} \Delta P_{ew} \\ \Delta Q_{ew} \end{bmatrix} = \begin{bmatrix} C_1 \Delta Z \\ C_2 \Delta Z \end{bmatrix} + \begin{bmatrix} D_1 |\Delta V_n| \\ D_2 |\Delta V_n| \end{bmatrix}
\tag{4.41}
$$

Details of C_1, C_2, D_1, and D_2 can be found in [24].

By expanding the network equations written generally written by (4.18), (4.42), and (4.43) can be obtained:

$$
P_{ej} - \mathrm{Re} \left\{ \sum_{k=1,\ k\neq j}^{N} V_{jk} I_{jk}^* \right\} - V_j^2 G_j = 0
\tag{4.42}
$$

$$
Q_{ej} - \mathrm{Im} \left\{ \sum_{k=1,\ k\neq j}^{N} V_{jk} I_{jk}^* \right\} - V_j^2 B_j + V_j^2 \sum_{k=1,\ k\neq j}^{N} \frac{B_{Ljk}}{2} = 0
\tag{4.43}
$$

where P_{ej} and Q_{ej} are active and reactive power injections at bus j, G_j and B_j are load conductance and susceptance at bus j, B_{Ljk} is the line charging connecting buses j and k, and N is the number of buses connected to bus j.

By linearizing (4.42) and (4.43) around an operating point and writing the small signal power flow equations at SG and wind generator buses separately, the system equations for a linearized Kron reduced wind-integrated power system model assuming zero input mechanical power deviation ($\Delta P_m = 0$) can be written by (4.44):

$$
\begin{bmatrix} M \Delta \ddot{\delta} \\ \Delta \dot{Z} \end{bmatrix} = A_M \begin{bmatrix} \Delta \delta \\ \Delta Z \end{bmatrix}
\tag{4.44}
$$

Details of the matrix compositions of A_M can be found in [24].

The expression given by (4.44) is similar to an SG only power system expression given by (4.27). Therefore, two-time scale analysis can be carried out for the wind-integrated system model similar to steps (4.31)–(4.39). This analysis is skipped for brevity and can be found in [24–26].

Identification of coherent groups is much more effective using data-driven techniques, which is explained in the section compared to analytical techniques. However, the significance of analytical techniques is that they do not depend upon the fault location. Furthermore, for an understanding of electrical features causing different coherent groups cannot be explored using data-driven techniques. The next section discusses the reduction of generator buses in a coherent group.

4.2.4.1.2 Generator bus aggregation (Zhukov's method)

In the following sections, classical methods for the aggregation of generator buses are reviewed. Zhukov's method of generator bus aggregation suggests that a set of PV buses can be reduced to a single equivalent generator bus by maintaining the current balance in the remaining nodes through ideal transformers [27–29]. This step is essential concerning the equivalent generator identification in the

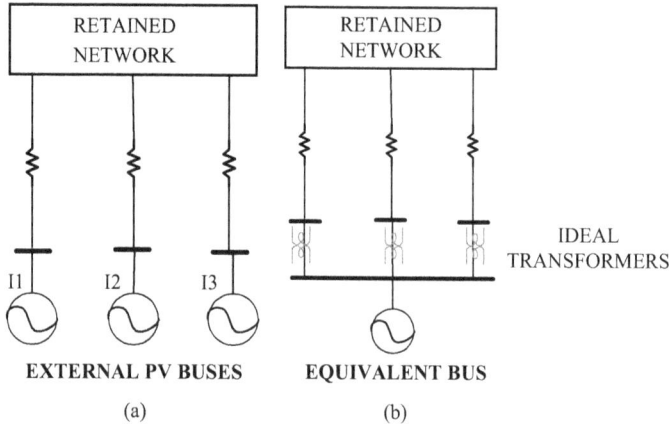

*Figure 4.8   Aggregation of coherent generator buses using Zhukov's technique :
(a) original network (b) aggregated network*

coherency-based reduction. Let us take a system with 3 PV buses as shown in Figure 4.8(a).

Current balance equations can be written by (4.45) and (4.46) for Figure 4.4(a) and 4.4(b), respectively:

$$\begin{bmatrix} I_r \\ I_g \end{bmatrix} = \begin{bmatrix} Y_{rr} & Y_{rg} \\ Y_{gr} & Y_{gg} \end{bmatrix} \begin{bmatrix} V_r \\ V_g \end{bmatrix} \tag{4.45}$$

$$\begin{bmatrix} I_r \\ I_G \end{bmatrix} = \begin{bmatrix} Y_{rr} & Y_{rG} \\ Y_{Gr} & Y_{GG} \end{bmatrix} \begin{bmatrix} V_r \\ V_G \end{bmatrix} \tag{4.46}$$

Since the reduction of the generator buses must not change any current injections to the retained network, the following expression can be derived by equalizing the value of I_r in two systems:

$$Y_{rr}V_r + Y_{rg}V_g = Y_{rr}V_r + Y_{rG}V_G \tag{4.47}$$

From (4.47), the following ratio satisfies a reduced system:

$$\frac{V_G}{V_g} = \frac{Y_{rg}}{Y_{rG}} = \gamma \tag{4.48}$$

The value γ is called the ideal transformer turns ratio.

Therefore, the reduced system defined by (4.46) can be rewritten using (4.49):

$$\begin{bmatrix} I_r \\ I_G \end{bmatrix} = \begin{bmatrix} Y_{rr} & \dfrac{Y_{rg}}{\gamma} \\ \dfrac{Y_{gr}}{\gamma} & Y_{GG} \end{bmatrix} \begin{bmatrix} V_r \\ \gamma V_g \end{bmatrix} \tag{4.49}$$

In (4.49), $\frac{Y_{rg}}{\gamma} = \frac{Y_{gr}}{\gamma}$ is satisfied since the Y matrix is symmetrical.

The value of γ is valid for a fixed power system state given that the values of voltages V_G and V_g remain constant. This does not mean that Zhukov's technique cannot be used in dynamic situations. The definition of coherency states that coherent generators exhibit a constant complex voltage ratio in their terminal buses during dynamics conditions [21]. Therefore, a group of coherent generators can be aggregated to one equivalent bus for dynamics equivalent representation. Further explanations are given for the application of Zhukov's technique for coherent bus reduction in [30]. The next step in the coherency-based reduction is the equivalent generator representation of the coherent group of machines.

Equivalent generator representation

The procedure explained in this step is for identifying equivalent generator parameters representing the dynamics of all generators in one coherent group. The classical equivalent generator parameter estimation can be mainly divided into three categories as follows [31]:

1. Frequency-weighted reduction
2. Frequency-weighted time domain method
3. Structure preservation technique

Frequency-weighted reduction

The aggregation of equivalent generator parameters from the frequency-weighted reduction method was introduced by the EPRI project conducted in the 1970s [32]. In this technique, the equivalent generator parameters are identified separately in terms of governor parameters, exciter parameters, synchronous machine parameters, power system stabilizer parameters for a group of coherent machines. Also, if there are different kinds of machines in a single coherent group such as steam turbine generators and hydro turbine generators, they are represented separately by equivalent generator models. The technique introduced in that paper is called a least square fit of the transfer functions in the frequency domain. To implement a least square fit, the first a generic model of the equivalent generator is obtained. For example, the transfer function for the equivalent governor of a hydro turbine system is expressed as

$$G_G^*(s) = \frac{PGV(s)}{\Delta\omega(s)} = K\frac{(1+sT_2)}{(1+sT_1)(1+sT_3)^2} \tag{4.50}$$

where K, T_1, T_2, and T_3 are the parameters to be estimated through the least square fit. The transfer function of an equivalent turbine of a hydro turbine system is expressed as

$$G_T^*(s) = \frac{P_M(s)}{PGV(s)} = \frac{1 - sT_w}{1 + \left(\frac{sT_w}{2}\right)} \tag{4.51}$$

where, T_w is the parameter to be estimated through the least square fit.

Since the governor and turbine are considered as a single system, the transfer functions $G_G^*(s)$ and $G_T^*(s)$ can be expressed in series to form the transfer function of the governor turbine system:

$$G_j(s) = G_G^*(s) \times G_T^*(s) \tag{4.52}$$

$$K \frac{(1 + sT_2)}{(1 + sT_1)(1 + sT_3)^2} \cdot \frac{1 - sT_w}{1 + \left(\frac{sT_w}{2}\right)} \tag{4.53}$$

The expression given by (4.52) is compared against the aggregated model of the governor turbine system given by (4.54) in the least square fit algorithm:

$$\sum G_j(s) = \sum \frac{\Delta P_{mj}(s)}{\Delta \omega(s)} \tag{4.54}$$

The least square fit for $G_j(s)$ and $\sum G_j(s)$ given by (4.52) and (4.54) is performed by selecting a set of discrete frequencies.

Frequency-weighted time domain method
The frequency-weighted time domain methods are primarily based on the balanced realization methods defined in control theory [33–35]. For a large dynamical system, the state-space representation could be difficult for both the observability and the reachability. Therefore, the balanced realization method uses a transformation of the state space model to reduce the degree of difficulty. To illustrate this, let us look at a linear time-invariant (LTI) system as follows [36]:

$$\dot{x} = Ax + Bu \tag{4.55}$$

$$y = Cx + Du \tag{4.56}$$

where x is the state of the system, u is the system inputs, and y is the system outputs. $A, B, C,$ and D are matrices where A is the system dynamic matrix of $A \in R^{n \times n}$, B is the system input matrix of $B \in R^{n \times n}$, $C \in R^{p \times n}$, $D \in R^{p \times m}$, n is the order of the system, m is the number of input variables, and p is the number of output variables.

Controllability (W_c) and observability (W_o) Gramians of the system given by (4.55) and (4.56) can be rewritten as follows:

$$W_o = \int_0^\infty e^{A^T t} C^T C e^{At} dt \tag{4.57}$$

$$W_c = \int_0^\infty e^{At} BB^T e^{A^T t} dt \tag{4.58}$$

where W_c defines the input-to-state behavior and W_o defines the state-to-output behavior.

These Gramians are also the solutions of the Lyapunov equations given by

$$AW_c + W_c A' + BB' = 0 \tag{4.59}$$

$$A' W_o + W_o A + C' C = 0 \tag{4.60}$$

There could be a similarity transformation for the state–space system such that both Gramians of the system will be equal to Hankel singular values:

$$W_{cb} = W_{ob} = \sum \tag{4.61}$$

where $\sum$ denotes Hankel singular values of the form

$$\sum = diag\{a_1, a_2 \ldots .a_n\} \tag{4.62}$$

$$\sigma_i = \sqrt{\lambda_i(W_c W_o)} = \sqrt{\lambda_i(W_o W_c)} \tag{4.63}$$

where λ_i are the eigenvalues of the system.

From the similarity transformation of the state space system given by (4.61) and the singular value decomposition technique, the following relationship can be obtained for the original system [37]:

$$W_c W_o = U \sum V^T \tag{4.64}$$

where U and V are left and right singular vectors with orthonormal properties, i.e., $UU^T = I$ and $VV^T = I$. Since U and V are orthonormal, (4.64) can also be written as

$$W_c W_o = U^T \sum V \tag{4.65}$$

Therefore, a transformation matrix T can be defined for the balanced system as follows:

$$T = W_c U \sqrt{\sum} \tag{4.66}$$

When the transformation matrix is applied to the original state space system given by (4.55) and (4.56), the balanced system can be obtained as follows:

$$\dot{x}_b = \tilde{A} x_b + \tilde{B} u \tag{4.67}$$

$$y = \tilde{C} x_b + Du \tag{4.68}$$

where $\tilde{A} = T^{-1}AT$, $\tilde{B} = T^{-1}B$, $\tilde{C} = CT$.

In the balanced system defined by (4.67) and (4.68), if the lowest singular values are truncated, the balanced truncated system or the reduced-order system can be defined as

$$\dot{x}_r = A_r x_r + B_r u \tag{4.69}$$

$$y = C_r x_r + D_r \tag{4.70}$$

The expression for the frequency-weighted time domain technique is defined by the minimization of the error function between the original system and the reduced system in time domain calculations. This comparison is given by (4.71):

$$e_r = \sqrt{\frac{\int_0^\infty \left(y_{org} - y_r\right)^2 dt}{\int_0^\infty \left(y_{org}\right)^2 dt}} \qquad (4.71)$$

where y_{org} is the original system and y_r is the reduced system.

The drawback of the frequency weighted method is that it necessitates at least a first-order representation of the equivalent system model [35]. However, original details of the generator models are typically given as the block diagram. Therefore, it requires an additional step of deriving the transfer function equations.

Structure preservation technique
The structure preservation technique involves the aggregation of coherent generator parameters in the time domain. Time domain methods are straightforward and hence can be used for online applications [38]. Since all the coherent generators are connected parallel to the equivalent generator bus as shown in Figure 4.4(b), it is reasonable to assume that there should be no change in the current injection before and after the equalization. The sum of apparent power injections from all generators in a coherent group must be equal to the apparent power injection from the equivalent generator as given (4.72):

$$S_e = \sum_{i=1}^{n} S_i \qquad (4.72)$$

Since generators are represented in parallel in Figure 4.8 (b), equivalent generator parameters for sub-transient reactance (x_d'', x_q''), transient reactance (x_d', x_q'), and steady-state reactance (x_d, x_q) can be approximately computed by the parallel reactance rule as follows:

$$\frac{1}{X_{eq}} = \frac{1}{X_1} + \frac{1}{X_2} + \frac{1}{X_3} \cdots\cdots\cdots \frac{1}{X_n} \qquad (4.73)$$

where X represents one reactance of $x_d'', x_q'', x_d', x_q', x_d, x_q$ and n is the number of machines in the coherent group.

This approximation of the equivalent impedance is made by disregarding the rotor angles of the coherent machines. However, a precise calculation of the equivalent impedances must be performed by considering rotor angles of the machines as given in [38].

Inertia and damping coefficients of the equivalent machine can be calculated by the series addition of those values of the coherent machines:

$$H_e = \sum_{i=1}^{n} H_i \qquad (4.74)$$

$$D_e = \sum_{i=1}^{n} D_i \qquad (4.75)$$

where n is the number of coherent machines.

Aggregation of time constants, exciter parameters, turbine governor system parameters are further given in [38,39].

4.2.4.2   Power system model-based techniques

The expression of the power system model given by (4.57) and (4.58) can be reduced by applying model reduction theories derived from linear control theory [40]. Dimensionality reduction has been a broadly discussed topic in the field of control theory for many years. The basic idea behind it is that in any dynamical system, around 90% of the dynamics are determined by the first few dominant modes of the system. Therefore, techniques such as modal analysis and singular value decomposition technique were developed to identify these dominant modes and truncate them from the system. Since these techniques explicitly focus on the power system model, they are defined as power system model-based techniques.

Modal analysis

The modal analysis identifies dominant modes (eigenvectors and eigenvalues) of a power system by performing an eigenvalue decomposition technique in the dynamic matrix (A) of the power system model. Then a model reduction is performed by retaining these dominant modes while eliminating weak modes. In [41], the selection of dominant eigenvalues is discussed by introducing a dominance index. By applying the modal analysis to the dynamic matrix (A) of (4.55), the following expression is obtained:

$$Av_i = v_i\lambda_i \tag{4.76}$$

where v_i is the ith right eigenvector of the dynamic matrix A and λ_i is its ith eigenvalue.

Singular value decomposition (SVD) technique

The SVD technique is used to rank energy modes (singular values) of a dynamical system in the descending order and truncate them to preserve only its dominant energy modes [42].

By applying the SVD technique to the dynamic matrix (A) of (4.55), the following expression can be obtained [42]:

$$A = VSU^T = \sum_{i=1}^{2(n-1)} u_i s_i v_i^T \tag{4.77}$$

where U is the left singular vector, V is the right singular vector, S is the singular vector, n is the number of system buses, u_i, s_i, and v_i^T are the ith columns of matrices V, S, U^T. U and V are unitary matrices. $(UU^T = U^TU = identity,$ $VV^T = V^TV = identity)$

In [43], by performing the SVD of the ES, dominant swing modes are extracted as first "r" modes. Afterwards, a reference generator is assigned to each extracted swing mode from the subset of generator swing curves. If a particular reference generator swing curve matches a swing mode of the ES, then that generator is considered as a characteristic generator. Finally, the external system is reconstructed only with the characteristic generators.

4.2.5   *Limitations of classical reduction techniques*

Focusing explicitly on the details of the power system model is one of the major drawbacks in classical model reduction techniques when applying to online stability analysis. This is because the scale of modern power systems is too large and hence, they drain a tremendous amount of computational power and consume long time periods. Furthermore, constant fluctuations in the power system state necessitate repetitive analysis of the model reduction techniques. Therefore, the tedious processes required to reduce the model involved in classical techniques must not be repeated in online stability analysis. For example, power system model-based techniques are too abstract involving complex mathematical calculations associated with a dynamic matrix of the size $n \times n$. For another instance, in the coherency-based technique, in which the generator speeds must be calculated prior to coherency clustering, it imposes limitations mentioned previously on large power systems. In comparison, since the slow-coherency technique is also strictly based on the power system model, it inherits similar dimensionality issues to the power system model-based techniques. In addition to the dimensionality issues and changes to the operating state, there is a lack of information sharing among different areas of a large-interconnected power system. Thus, there must be techniques that do not rely solely upon the model details. This allowed room for measurement-based dynamic model reduction techniques, which are fueled by modern advancements in data engineering and massive installations of PMUs in the power systems.

4.2.6   *Research gaps and conclusions*

To assess the power system strength in all or some of the nodes of a bulk power system, there is a need to work out reduced-size equivalent models for the power system. This book chapter discussed the classical reduction techniques of the power system that primarily focus on the detailed model of the power system. From these techniques, Kron's technique, Ward's technique, and the REI technique are the most prominent static reduction techniques. All three of them rely on a fundamental principle in electrical engineering, i.e., Gaussian elimination technique that is applied for redistributing the currents in the remaining nodes when one node of the system is eliminated. However, this principle can only be applied to non-generator buses. If generator buses need to be eliminated, then it can only be performed considering generators as negative loads, which is known as the generator netting process. Consequently, the dynamics of the reduced system could be deviated from the unreduced system from the generator netting process.

The dynamic model reduction techniques incorporate two main streams, i.e., power system model-based reduction and coherency-based reduction. The former focuses on representing the power system as a state–space model and applying model reduction techniques to the dynamic matrix of that. The latter focuses on the coherent groups of the ES for a disturbance in the study system, which will ultimately be represented as equivalent generators. To be specific, groups of machines that oscillate together are considered as coherent groups, which will be represented as aggregated dynamics in the ES. Therefore, this technique further reduces generated buses in the ES while dynamically reducing its characteristics.

To conclude, all the classical reduction techniques start from a load flow solution and should be redone when the operating conditions change. The contemporary power systems that consist of a high penetration of IBGs, which are predominantly supplied by intermittent energy sources like wind and solar, the operating point is frequently changing. Therefore, new types of equivalent techniques that can adapt to the changes in the operating point could be highly useful. Furthermore, the accuracy of the power flows at the tie lines should be corrected for accurate system strength studies in the SS. By focusing on these two aspects, two of the main challenges in modern power systems, i.e., reduction time and accuracy can be evaded.

References

[1] S. R. Sinsel, R. L. Riemke, and V. H. Hoffmann, "Challenges and solution technologies for the integration of variable renewable energy sources—a review," *Renewable Energy*, vol. 145, pp. 2271–2285, 2020.

[2] R. Yan, T. K. Saha, F. Bai, and H. Gu, "The anatomy of the 2016 South Australia blackout: a catastrophic event in a high renewable network," *IEEE Transactions on Power Systems*, vol. 33, no. 5, pp. 5374–5388, 2018.

[3] A. E. M. Operator, "System strength in the NEM explained," in *System Strength*, 2020.

[4] L. Yu, H. Sun, S. Xu, B. Zhao, and J. Zhang, "Critical system strength evaluation of the power system with high penetration of renewable energy generations," *CSEE Journal of Power and Energy Systems*, vol. 8, pp. 710–720, 2021.

[5] D. Wu, G. Li, M. Javadi, A. M. Malyscheff, M. Hong, and J. N. Jiang, "Assessing impact of renewable energy integration on system strength using site-dependent short circuit ratio," *IEEE Transactions on Sustainable Energy*, vol. 9, no. 3, pp. 1072–1080, 2017.

[6] B. de Metz-Noblat, F. Dumas, and C. Poulain, "Calculation of short-circuit currents," *Cahier Technique*, vol. 158, 2005.

[7] SIEMENS, PSSE Program Application Guide, Volume 1, 2017.

[8] W. Yan, C. Zhang, J. Tang, W. Qian, S. Li, and Q. Wang, "A black-box external equivalent method using tie-line power mutation," *IEEE Access*, vol. 7, pp. 11997–12005, 2019.

[9] P. S. Kundur and O. P. Malik, "General characteristics of modern power systems: structure of the power system," in *Power System Stability and Control*. New York, NY: McGraw-Hill Education, 2022.

[10] J. Machowski, Z. Lubosny, J. W. Bialek, and J. R. Bumby, *Power System Dynamics: Stability and Control*. New York, NY: John Wiley & Sons, 2020.

[11] M. Gibbard and D. Vowles, "IEEE PES task force on benchmark systems for stability controls simplified 14-generator model of the south east Australian power system," in *The University of Adelaide*, vol. 5005, 2014.

[12] G. Kron, "Tensor analysis of networks," *New York Times*, 1939.

[13] Q. Ploussard, L. Olmos, and A. Ramos, "An efficient network reduction method for transmission expansion planning using multicut problem and Kron reduction," *IEEE Transactions on Power Systems*, vol. 33, no. 6, pp. 6120–6130, 2018.

[14] J. B. Ward, "Equivalent circuits for power-flow studies," *Electrical Engineering*, vol. 68, no. 9, pp. 794–794, 1949.

[15] R. Van Amerongen and H. Van Meeteren, "A generalised ward equivalent for security analysis," *IEEE Transactions on Power Apparatus and Systems*, vol. 6, pp. 1519–1526, 1982.

[16] P. Dimo, *Nodal Analysis of Power Systems*, UK: Abacus Press, 1975.

[17] X.-F. Wang, Y. Song, and M. Irving, "Mathematical model and solution of electric network," in *Modern Power Systems Analysis*. New York, NY: Springer Science & Business Media, 2010.

[18] P. Kundur, J. Paserba, V. Ajjarapu, *et al.*, "Definition and classification of power system stability IEEE/CIGRE joint task force on stability terms and definitions," *IEEE Transactions on Power Systems*, vol. 19, no. 3, pp. 1387–1401, 2004.

[19] J. H. Chow, *Power System Coherency and Model Reduction*. New York, NY: Springer, 2013.

[20] J. Chow, J. Winkelman, M. Pai, and P. Sauer, "Singular perturbation analysis of large-scale power systems," *International Journal of Electrical Power & Energy Systems*, vol. 12, pp. 117–126, 1990.

[21] R. Podmore, 'Identification of coherent generators for dynamic equivalents," *IEEE Transactions on Power Apparatus and Systems*, no. 4, pp. 1344–1354, 1978.

[22] I. Tyuryukanov, M. Popov, M. A. van der Meijden, and V. Terzija, "Slow coherency identification and power system dynamic model reduction by using orthogonal structure of electromechanical eigenvectors," *IEEE Transactions on Power Systems*, vol. 36, no. 2, pp. 1482–1492, 2020.

[23] J. H. Chow, R. Galarza, P. Accari, and W. W. Price, "Inertial and slow coherency aggregation algorithms for power system dynamic model reduction," *IEEE Transactions on Power Systems*, vol. 10, no. 2, pp. 680–685, 1995.

[24] S. Chandra, D. F. Gayme, and A. Chakrabortty, "Time-scale modeling of wind-integrated power systems," *IEEE Transactions on Power Systems*, vol. 31, no. 6, pp. 4712–4721, 2016.

[25] S. Chandra, M. D. Weiss, A. Chakrabortty, and D. F. Gayme, "Impact analysis of wind power injection on time-scale separation of power system oscillations," in *2014 IEEE PES General Meeting/Conference & Exposition*, 2014. New York, NY: IEEE, pp. 1–5.

[26] S. Mukherjee, A. Chakrabortty, and S. Babaei, "Modeling and quantifying the impact of wind penetration on slow coherency of power systems," *IEEE Transactions on Power Systems*, vol. 36, no. 2, pp. 1002–1012, 2020.

[27] J. Machowski, A. Cichy, F. Gubina, and P. Omahen, "External subsystem equivalent model for steady-state and dynamic security assessment," *IEEE Transactions on Power Systems*, vol. 3, no. 4, pp. 1456–1463, 1988.

[28] J. Machowski, Z. Lubosny, J. W. Bialek, and J. R. Bumby, "Power system model reduction – equivalents," in *Power System Dynamics: Stability and Control*. New York, NY: John Wiley & Sons, 2020.

[29] J. Machowski, F. Gubina, and P. Omahen, "Power system transient stability studies by Lyapunov method using coherency based aggregation," in *Power Systems and Power Plant Control*. New York, NY: Elsevier, 1987, pp. 239–244.

[30] J. H. Chow, "Coherency in power systems," in *Power System Coherency and Model Reduction*, vol. 84. New York, NY: Springer, 2013.

[31] R. Singh, M. Elizondo, and S. Lu, "A review of dynamic generator reduction methods for transient stability studies," in *2011 IEEE Power and Energy Society General Meeting*, 2011. New York, NY: IEEE, pp. 1–8.

[32] A. J. Germond and R. Podmore, "Dynamic aggregation of generating unit models," *IEEE Transactions on Power Apparatus and Systems*, vol. PAS-97, pp. 1060–1069, 1978.

[33] U. M. Al-Saggaf and G. F. Franklin, "Model reduction via balanced realizations: an extension and frequency weighting techniques," *IEEE Transactions on Automatic Control*, vol. 33, no. 7, pp. 687–692, 1988.

[34] C.-A. Lin and T.-Y. Chiu, "Model-reduction via frequency weighted balanced realization," *Control-Theory and Advanced Technology*, vol. 8, no. 2, pp. 341–351, 1992.

[35] B.-K. Choi, H.-D. Chiang, H. Wu, H. Li, and C. Y. David, "Exciter model reduction and validation for large-scale power system dynamic security assessment," in *2008 IEEE Power and Energy Society General Meeting-Conversion and Delivery of Electrical Energy in the 21st Century*, 2008. New York, NY: IEEE, pp. 1–7.

[36] A. Ramirez, A. Mehrizi-Sani, D. Hussein, *et al.*, "Application of balanced realizations for model-order reduction of dynamic power system equivalents," *IEEE Transactions on Power Delivery*, vol. 31, no. 5, pp. 2304–2312, 2015.

[37] A. G. Akritas and G. I. Malaschonok, "Applications of singular-value decomposition (SVD)," *Mathematics and Computers in Simulation*, vol. 67, no. 1–2, pp. 15–31, 2004.

[38] M. L. Ourari, L.-A. Dessaint, and V.-Q. Do, "Dynamic equivalent modeling of large power systems using structure preservation technique," *IEEE Transactions on Power Systems*, vol. 21, no. 3, pp. 1284–1295, 2006.

[39] R. Guo, "Dynamic system equivalents using integrated PSS/E and Python for transient stability studies," University of Manitoba, Department of Electrical and Computer Engineering, Master of Science Thesis, 2021.

[40]   S. D. Ðukić and A. T. Sarić, "Dynamic model reduction: an overview of available techniques with application to power systems," *Serbian Journal of Electrical Engineering*, vol. 9, no. 2, pp. 131–169, 2012.

[41]   L. Litz, "Order reduction of linear state-space models via optimal approximation of the nondominant modes," *IFAC Proceedings Volumes*, vol. 13, no. 6, pp. 195–202, 1980.

[42]   G. W. Stewart, "On the early history of the singular value decomposition," *SIAM Review*, vol. 35, pp. 551–566, 1993.

[43]   S. Wang, S. Lu, N. Zhou, G. Lin, M. Elizondo, and M. Pai, "Dynamic-feature extraction, attribution, and reconstruction (DEAR) method for power system model reduction," *IEEE Transactions on Power Systems*, vol. 29, no. 5, pp. 2049–2059, 2014.

Chapter 5

Dynamic model reduction of IBRs-rich power networks for fast assessment of power system strength – part 2: data-driven techniques

Lahiru Aththanayake Adhikarige[1], Ameen Gargoom[1] and Nasser Hosseinzadeh[1]

Abstract

This chapter is an extension of Chapter 4. It introduces new techniques for reducing the size of modern large power systems including data-driven techniques, and measurement-based techniques. It also includes reduction techniques for wind power plants, solar power plants, microgrids, and active distribution networks (ADNs). Also, a case study is presented on dynamic model reduction of a power system using LSTM recurrent neural networks. The chapter concludes that the best way of reducing a large power system model is a combination of classical and data-driven techniques. This is because in most cases, some data of the system is available for the model, and only a few dynamic data are unknown such as generator dynamics of only some generating units, for which data-driven techniques can be used to compute them approximately. In the parameter estimation process, a set of critical parameters should be selected based on the sensitivity analysis to decrease the burden as well as reaching the local optimum. In this way, a robust model of the external system, i.e., all parts of the system except the area under study can be implemented for system strength assessment studies. A case study is performed on a simplified Australian 14 generator model to show the potential of measurement-based techniques in the reduction of power systems and their strengths limitations and limitations in applying to modern IBRs-rich power networks are examined.

5.1 Data-driven techniques

In the past two decades, the field of data engineering has undergone a massive transformation with new types of machine learning-based techniques proposed to extract information from datasets and derive meaningful interpretations. The field

---

[1]Centre for Smart Power and Energy Research, School of Engineering, Deakin University, Australia

of machine learning is continuously evolving and a brief overview of current trends in machine learning can be found in [1,2]. This revolution in the big data field, together with massive installations of phasor measurement units (PMUs) in the power systems, accelerated the application of data-driven techniques to solve problems in power systems. For example, deep artificial neural networks (DNNs) have been deployed for load forecasting [3], solar and wind power predictions [4], smart grid applications [5], and transient stability assessment [6]. The deep recurrent neural networks (RNNs) are promising for capturing sequence types of dynamics in the power systems. Therefore, they have been deployed for designing power system stabilizers [7], load forecasting [8], wind power forecasting [9], solar power forecasting [10], identification of power system fluctuations [11], fault detection [12], and cyber-attacks in smart grids [13]. Novel clustering techniques can also be applied to solve many problems associated with power systems. For example, wind power predictions [14,15] and solar power predictions [16] are achieved through clustering algorithms. There are many more applications of machine learning techniques for power system applications. Therefore, the rest of the discussion will only be focused on the application of data-driven techniques to predict the dynamics of an ES system in system strength assessment (SSA).

The application of data-driven techniques for SSA focuses on reducing the size of an ES of a power system and represents it as a reduced order model to demonstrate the approximate dynamics of the original. The size reduction of the ES using measurement data can be looked at in two ways, i.e., black-box and grey-box approaches [17]. In the former, only the measurement data at the boundary nodes between the SS and ES are known. Therefore, the ES dynamics should be estimated either through a machine learning technique or by identifying an estimated model of the ES through a parameter estimation technique. In the grey-box technique, some details of the structure of the ES are known but the rest needs to be approximated through measurement data.

Before understanding how data-driven techniques can be applied to power system dynamic model reduction problem, it is important to understand mathematical concepts behind a dynamic model reduction of the ES of a power system. The power system can be identified as a system with a set of first-order nonlinear ordinary differential equations and a set of algebraic equations as given by (5.1) and (5.2):

$$\dot{x} = f(x, u, \ t) \tag{5.1}$$

$$y = g(x, u) \tag{5.2}$$

where x is the state vector, u is the input vector y is the output vector, and t denotes the time that relates the state vector to its derivate.

Equations (5.1) and (5.2) can be separately written for the SS and the ES in the following way:

$$\dot{x}_s = f_s(x_s, u_s, \ t) \ , \ y_s = g_s(x_s, x_e, \ u_s, \ u_e, t) \tag{5.3}$$

$$\dot{x}_e = f_e(x_e, u_e, \ t) \ , \ y_e = g_e(x_s, x_e, \ u_s, \ u_e, t) \tag{5.4}$$

where s denotes the SS and e denotes the ES.

Equations (5.3) and (5.4) emphasize the interconnection between the ES and the SS since their algebraic equations are decoupled with the other network. This is the main reason why it is mandatory to include the ES dynamics in any SSA of the SS. However, due to computation constraints, time constraints, and other limitations (discussed in the previous chapter) of using the full ES model in SSA simulations, reduced order models are proposed for the ES. The mathematical representation of the power system separated into SS and ES considering a reduced order model for the ES is by (5.5) and (5.6):

$$\dot{x}_s = f_s(x_s, u_s, \ t) \ , \ y_s = g_s(x_s, \overline{x_e}, \ u_s, \ \overline{u_e}, t) \tag{5.5}$$

$$\overline{\dot{x}_e} = \overline{f_e}(\overline{x_e}, \overline{u_e}, \ t) \ , \ \overline{y_e} = \overline{g_e}(x_s, \overline{x_e}, \ u_s, \ \overline{u_e}, t) \tag{5.6}$$

where $\overline{x_e}$ is the state vector of the reduced system, $\overline{f_e}$ is the dynamic equations of the reduced ES, $\overline{g_e}$ is the algebraic equations of the reduced ES, and $\overline{u_e}$ is the input vector of the reduced ES.

The primary focus of any dynamic model reduction technique is to find the lowest order reduced equivalent system given by (5.6) that also predicts the best approximated dynamics of the ES given by (5.4).

5.2 Black-box identification of the ES

The black-box approach to predict the ES dynamics only has the access to the boundary conditions. These measurement data include boundary bus properties such as voltage (V_s), voltage angle (δ_s), and frequency deviation (f_{dev}) and tie line properties such as active power flow (P) and reactive power flow (Q). In this field, machine learning techniques are widely deployed to predict the tie line power flows from the boundary bus properties. The following diagram shows the black-box representation of the ES in the simplified Australian 14 generator system (AU14G) considering the Queensland system as the test system.

As seen from Figure 5.1, the ES of the AU14 model that includes NSW, VIC, TAS, and SA is reduced to a simple black-box model. This black-box model necessarily does not need to be a physical electrical system. For example, a DNN or an RNN can be trained to predict the dynamics of the ES as a black-box model. Another way of representing the ES as a black-box model is identifying the parameters of a differential algebraic system, a transfer function model. In this section, only the data-driven techniques applied for the model reduction at the transmission level are described. Other types of data-driven equivalent techniques applied for dynamic model reduction of wind power plants, solar power plants, microgrids, and active distribution networks (ADNs) are discussed in Section 5.3.

5.2.1 Non-parametric techniques

In [18], a combination of a feed-forward artificial neural network (ANN) and a recurrent neural network (RNN) is used to represent the ES. Specifically, a "bottleneck" ANN is used for state extraction and state reduction, and the RNN is used

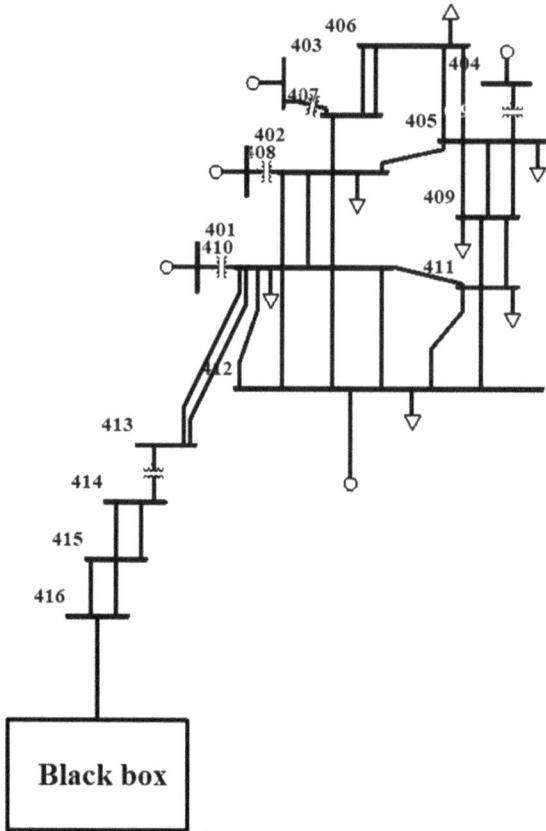

Figure 5.1   Black-box equivalent of the ES model

for state prediction of the ES. A "bottleneck" ANN refers to a structure in which two hidden layers are compressed since the output of one hidden layer becomes the input to the successive hidden layer [19] (in this model, the output of the mapping layer is the input to the demapping layer). The equivalent circuit obtained from the proposed technique was able to predict both stable and unstable responses. It is also discovered that the mean squared error (MSE) of the predicted result with the actual result varies with the number of neurons in the "bottleneck layer" and the optimum number of neurons refers to the minimum MSE. It is common in ANNs to show inaccurate results when the response of the tested fault has largely deviated from the training dataset. In this case, when the clearing time of the tested fault is larger than the training faults, errors were observed in the response from the ANN. In [20], an artificial neuro-fuzzy system (ANFIS) is identified for the ES dynamic model from the boundary measurements. An ANFIS is a combination of neural networks and fuzzy inference systems working together for predicting highly nonlinear dynamical systems [21]. In [22], the ES is replaced with a dynamic algebraic neural network. It emphasizes the importance of simultaneously solving dynamic algebraic equations of the SS and the ES for precise identification of the ES. Dynamic equations of the reduced dynamic algebraic equivalent are identified in the discrete-time domain first and then converted to the continuous-time domain. Full ANN consists of four layers with two layers for the state-space model and two layers for the algebraic model. For simplicity, the number of state variables is set to five and it is corresponding to the number of neurons in the second layer (output layer of the state-space model). Phasor values of voltage and currents of boundary nodes are considered as inputs to the ANN meanwhile magnitudes of voltage and current waveforms of boundary nodes are predicted from the ANN. The lack of physical interpretability in earlier ANN-based techniques is partially addressed in this technique since the ANN is used here to estimate the parameters of the dynamic algebraic model. However, rather than simply modeling the ES architecture as a dynamic algebraic equivalent, it would make more sense to determine components of the ES such as equivalent generators and loads.

5.2.2   Parameter estimation techniques

The identification of a reduced order model for the ES through the boundary mea-surement data involves parameter estimation techniques, which is also known as the hyperparameter optimization techniques. There are two main categories of hyper-parameter optimization in the literature, i.e., analytical techniques and metaheuristic techniques [23]. The analytical techniques are also called deterministic techniques and the metaheuristic techniques are also called stochastic techniques. Newton Raphson [24] and Gauss-Seidel [25] methods are two examples of analytical techniques while Genetic Algorithm (GA) [26] and Particle Swarm Optimization [27] are widely pop-ular metaheuristic techniques. The key difference between analytical and metaheuristic techniques is that the analytical approach relies upon test results and a numerical technique to solve the equations of the system while minimization of an objective function like (5.7) is used for metaheuristic techniques:

$$x^* = \arg\min_{x \in X} f(x) \tag{5.7}$$

where x^* is the hyperparameter to be estimated, $f(x)$ is called the objective function of the minimization algorithm, and X is called the hyperparameter search space.

Metaheuristic techniques utilize exploration and exploitation properties when achieving the targets [28]. Exploration suggests the ability to find the solutions while exploitation suggests the ability to optimize the search space to find those solutions. Metaheuristic optimization functions need not only be focusing on single objective functions as given by (5.7). The objective function given in (5.8) can be expanded to show multi-objective optimization as follows:

$$F(x) = (f_1(x), f_2(x) \ldots \ldots f_k(x)) \tag{5.8}$$

where k is the number of objective functions.

There are two main metaheuristic optimization techniques i.e. biologically inspired (evolutionary) [29] and physics-inspired algorithms [30]. Evolutionary algorithms such as optimization GA, PSO, Bee Colony, Ant colony, Grey wolf optimization, Bat algorithm, and Crow search algorithms are inspired by natural techniques to achieve a target. Physics-based techniques such as simulated annealing, gravitational search, quantum mechanics-based algorithms, universe theory-based algorithms, electrostatics-based algorithms, and chaotic optimization are based on physical phenomenon to achieve a target.

In [31], the ES is replaced with one equivalent generator model, and its parameters are determined through a neural network integrated with a radial basis function. A radial basis function is a mathematical technique to approximate a function with multiple variables. When it is embedded to an artificial neural network, it has proven to produce better accuracy than conventional ANNs for datasets consisting of lesser number of inputs [32]. Inputs to the neural networks are selected from measurement data in transient waveforms of the SS. These measurements are peak overshoot, decay constant, and period of oscillation of frequency and voltage variations. Output predictions from the neural network are effective inertia constant, damping coefficient, direct axis synchronous reactance, direct axis transient reactance, and open circuit field time constant of the equivalent generator model. An advantage of this technique over the classical techniques is that, rather than using a linear system model, nonlinear neural networks are used to identify equivalent generator parameters, which can effectively eliminate the linearization error. Additionally, since the prediction results for only five parameters are used, an intensive training is not required. Parameter estimation of a second-order synchronous generator model using a nonlinear least square technique is proposed in [33]. Although satisfactory results were obtained from one equivalent generator model in both papers, the dynamics of a large power system cannot be captured with a single equivalent generator model. Furthermore, it is not investigated how the equivalent model would cope with the changes in operating conditions of both the ES and the SS.

In [34], a dynamic equivalent technique is proposed for the ES using a modified REI technique combined with the Levenberg–Marquardt weighted least square nonlinear optimization technique [35]. The modified REI technique is used to determine the static equivalent part of the ES. For the dynamic part of the ES, parameters of three components, i.e., an equivalent synchronous generator model and a composite equivalent model are determined through the Levenberg–Marquardt technique. Identifying an ES consisting of both equivalent generators and equivalent load models with minimum information about the ES is an attractive method. However, the time consumed to identify the network is not significantly low in comparison to other available techniques. Additionally, identified synchronous generator model is not a detailed synchronous generator model with all components such as automatic voltage regulator. In the meantime, this technique could be expanded to identify other equivalent components of the ES such as equivalent inverter-based resources (IBRs).

A novel transfer function estimation-based technique is developed in [36], which can be used as a PSSE user-defined GNE model. In this technique, the ES is estimated by the following general transfer function model:

$$G(s) = K \frac{(s - z_1)(s - z_2) \ldots \ldots \ldots (s - z_{m-1})(s - z_m)}{(s - p_1)(s - p_2) \ldots \ldots \ldots (s - p_{n-1})(s - p_n)} \tag{5.9}$$

where z is a zero, m is the number of zeros, p is a pole, and n is the number of poles.

Since the transfer function used in (5.9) is derived from a single disturbance, it lacks performance for unprecedented disturbances in the SS. Therefore, this technique is enhanced by considering multiple faults in [37]. In this technique, a correlation coefficient is used to find the similarities in tie line power flows for multiple disturbances and they are grouped accordingly. Then, a comparison made between tie-line power flows for a model trained with single event and multiple events showcases significant improvements in the latter. A comparison of black-box identification of the ES using linear regression models, support vector regression and ANNs are given in [38].

5.3   Application of measurement-based techniques to IBR-integrated networks

The identification of coherent groups in SG-based power systems is performed by measurement data of internal rotor angle deviation in SGs. This is because the internal rotor angle deviation is directly synchronized with the oscillations in the network in the event of a disturbance. For systems with a low penetration of IBRs, the coherency identification is simple since IBRs-interconnected buses are typically regarded as negative loads and need not be considered in the coherency identification. However, for a high penetration level of IBRs in the system, the stability of the power system is affected by the dynamics behind IBRs as well. Therefore, the presence of IBRs must not be disregarded in the coherent clustering and identification of coherency in IBRs-interconnected buses are an essential part of the process. There are a few key challenges in this process as follows:

1. Rotor angle deviation signal cannot be considered as the measurement signal for coherency identification since IBRs are interconnected through a power electronic interface.
2. Coherent groups must be identified separately for IBRs and SGs.
3. Fast dynamics and reduced inertia in IBRs can affect the coherency of traditional SGs.
4. IBRs with a lower power capacity must be disregarded in coherency identification since there can be many smaller scale IBRs connected to the network. These smaller scale IBRs can be represented with negative loads later in the network reduction process.

A solution to the first problem mentioned above is proposed in [39] by considering the frequency deviation signals of the terminal buses of all generators for coherency identification. This technique is analogous to finding coherent groups using rotor angle deviation signals in SGs only power systems. The identification of coherent clusters using measurement-based signals will be discussed in the next section.

5.3.1   Measurement-based coherency identification

This section comprises of data-driven clustering techniques for the identification of coherent generators in the power systems using measurement data. In this section, data-driven coherent clustering techniques used for both IBR integrated and SGs only power networks are discussed. Even though the focus in this section is techniques of identifying coherent clusters in systems integrated with IBRs, the techniques applicable to SGs only power networks can be extended to systems with IBRs by considering terminal bus bar frequency deviation signal of generator buses instead of internal rotor angle deviation signal as explained previously. Therefore, the focus of this section is time-series clustering of measurement data. First, let us take two times series examples as given by (5.10) and (5.11):

$$F_1 = \{f_1, f_2, f_3, f_4 \ldots \ldots f_T\} \tag{5.10}$$

$$F_2 = \{f_1, f_2, f_3, f_4 \ldots \ldots f_T\} \tag{5.11}$$

where T is the number of samples in the time series.

According to [40], there are three basic ways to find the similarity between these two times series i.e., shape-based, feature-based, and model-based. In the shape-based approach, for example in dynamic time warping (DTW) technique, the clusters are explicitly identified based on the raw-data. Whereas in the feature-based techniques, for instance in the Pearson correlation coefficient method, raw time series data is converted into a feature vector before identifying the clusters. In the model-based method, raw data is converted into parameters of a model before applying a clustering technique.

Raw measurement data usually contain a lot of noise while being spanned from a higher dimension as well. Inserting such raw data into a clustering algorithm would drain a lot of computational power and inaccurate results. Therefore, feature extraction is used before the time series clustering that transforms the time series data at a higher dimension to a lower dimension. Feature extraction is an important step in time series clustering because of the following reasons:

- It reduces computational memory.
- It increases efficiency.
- It speeds up clustering.
- It increases accuracy.

Some common methods of feature extraction are as follows:

- Principal component analysis (PCA)
- Discrete Fourier transform
- Discrete wavelet transform
- Singular value decomposition (SVD)
- Adaptive Cosine piecewise constant approximation

Now let us look at the DTW technique, the PCC, the PCA, and the SVD in detail.

5.3.1.1   Dynamic time warping

The idea behind the DTW is finding an optimum path through a distance matrix (d) given by solving the minimization problem given by (5.12):

$$\text{DTW}(F_1, F_2) = \min_w \left\{ \left(\sqrt{\sum_{k=1}^{K} w_k} \right) \right\} \tag{5.12}$$

Where the elements w_k of the distance matrix are given by (5.13):

$$d(i,j) = (s_i - t_j)^2 \tag{5.13}$$

The DTW technique is highly effective for data involving shifts in the time series as the Euclidian distance (EU) technique fails to identify those clusters [41]. One drawback of the DTW technique is the computational power required to minimize the problem given by (5.12). This is verified by the drawbacks of shape-based clustering as mentioned in [40].

5.3.1.2 Pearson correlation coefficient

The term correlation is used to define the similarity or the association between two mathematical or statistical variables [42]. The value of correlation can vary between -1 and $+1$, -1 being inversely correlated, $+1$ being correlated and 0 being absolutely contrast. For time-series data, there are two main types of correlations, i.e., cross-correlation and auto-correlation [43]. Cross-correlation is the analysis of correlation between two series data. For example, when analyzing the correlation between the two time series given by (5.10) and (5.11), cross-correlation must be used. Auto-correlation is important when analyzing the dependency of its own data. However, when identifying the coherent groups, the auto-correlation has no significance and hence will not be discussed in this section.

The expression for the Pearson correlation index is given by expression (5.14) [44]:

$$\rho_{X,Y} = \frac{E(X - \mu_X)(Y - \mu_Y)}{\sigma_X \sigma_Y} \tag{5.14}$$

where X and Y are two times series having μ_X and μ_Y mean values and σ_X and σ_Y standard deviation values. A value of $+1$ or -1 for $\rho_{X,Y}$ indicates a perfect correlation while 0 indicates an absolute contrast.

5.3.1.3 PCA

The PCA is a dimensionality reduction technique that transforms a given dataset into orthogonal components (principal components) so that the correlation between the data becomes minimum. These principal components are selected in directions where the variance of the dataset is maximum. Eventually, the dataset will consist of information which is most significant and mutually uncorrelated. However, the PCA is a linear transformation of the original dataset into another base that can represent the dataset optimally.

Let us consider the mathematical aspects of the PCA. Let us take the original dataset given by a matrix X ($m \times n$) and the transformed dataset given by a matrix Y ($m \times n$). Then the linear transformation of X into Y can be represented by (5.15) [45]:

$$Y = PX \tag{5.15}$$

where P ($m \times m$) is the linear transformation matrix.

As explained previously, the PCA analysis selects the transformation matrix P such that the variance of the transformed dataset Y is aligned in the maximum variance directions of the original matrix X.

The variance for the time series F_1 given by (5.16) can be written as follows:

$$\sigma^2 = \sum \frac{(f_{1i} - \overline{f_1})}{n - 1} \tag{5.16}$$

where n is the number of discrete samples and $\overline{f}$ is the mean.

Then covariance for the series can be written as follows:

$$\sigma^2 = \frac{F_1 F_1^T}{n - 1} \tag{5.17}$$

The idea behind the PCA analysis is that that the data should be least correlated when transformed.

Now let us extend the covariance given by (5.17) for two series F_1 and F_2 as follows:

$$C_{F_1, F_2} = \frac{\sum (f_{1i} - \overline{f_1})(f_{2i} - \overline{f_2})}{n - 1} \tag{5.18}$$

If we extend the number of time series to m with n data points in each time series, then the input vector can be written as

$$X = \begin{bmatrix} x_{11} & x_{12} & \cdots & x_{1n} \\ x_{21} & x_{22} & \cdots & x_{2n} \\ . & . & & . \\ x_{m1} & x_{m2} & \cdots & x_{mn} \end{bmatrix} \tag{5.19}$$

Hence, we can construct the covariance matrix C_x for a $(m \times n)$ X matrix by calculating the covariance of X with its transpose X^T $(n \times m)$:

$$X^T = \begin{bmatrix} x_{11} & x_{21} & \cdots & x_{m1} \\ x_{12} & x_{22} & \cdots & x_{m2} \\ . & . & & . \\ x_{1n} & x_{2n} & \cdots & x_{mn} \end{bmatrix} \tag{5.20}$$

This is similar to two time series covariance given in (5.17) but m time series consisting of n datapoints each:

$$C_x = \frac{1}{n - 1} X X^T \tag{5.21}$$

where matrix C_x is the size of $m \times m$.

Since the definition of the PCA suggests that covariance should be minimal between the data paints, a linear transformation P is selected such that dynamics of the system is maximized while minimizing the noise.

5.4 Case study – identification of the ES coherent generators in the AU14G system using the dynamic time warping technique

In this case study, coherent groups of the ES are identified using the dynamic time warping (DTW) technique in the AU14G system. The AU14G is a benchmarked test system by IEEE task force known as "Simplified Australian 14 generator model" (AU14G) that replicates the full Australian system in simplified a version. This test system consists of 14 generators, 5 static VAR compensators, 59 buses, and 104 lines with voltage ranging from 15 kV to 500 kV, as shown in Figure 5.8 [46]. It should be noted that this system is a reduced equivalent system of the conventional Australian power system. It was developed before the advent of renewable energy sources penetration into the electrical power network, i.e., before the year 2014.

In the power system shown in Figure. 5.2, considering the QLD system as the SS, an identical three phase to ground fault was simulated at bus 405 of the QLD system at 0.2 s for 0.5 s. Following the fault, a transmission line connecting buses 405 and 406 was set out of service. After that, the post fault dynamics were carried out for 6 s. Consequently, the bus bar frequency deviations of all ES system generator buses were measured. For coherency identification, only the post transient oscillations are of interest. Therefore, the measurement data is truncated from 3 to 6 s to eliminate the transient periods during and immediately after the faults. These measurements are shown in Figure 5.3.

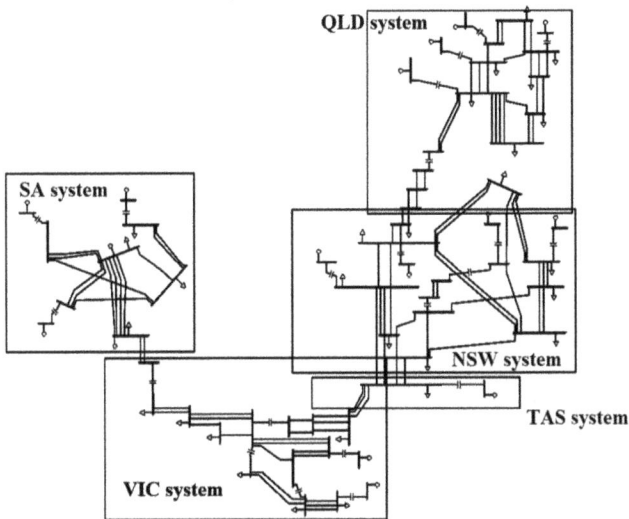

Figure 5.2 Simplified 14-Generator Model of the South-East Australian Power System – IEEE benchmark model

As seen in Figure 5.3, some generators tend to form coherent clusters. To identify these coherent groups, the measurement data shown in Figure 5.3 is fed into the DTW clustering algorithm. Then to find the optimal number of clusters, the elbow method is used as shown in Figure 5.4.

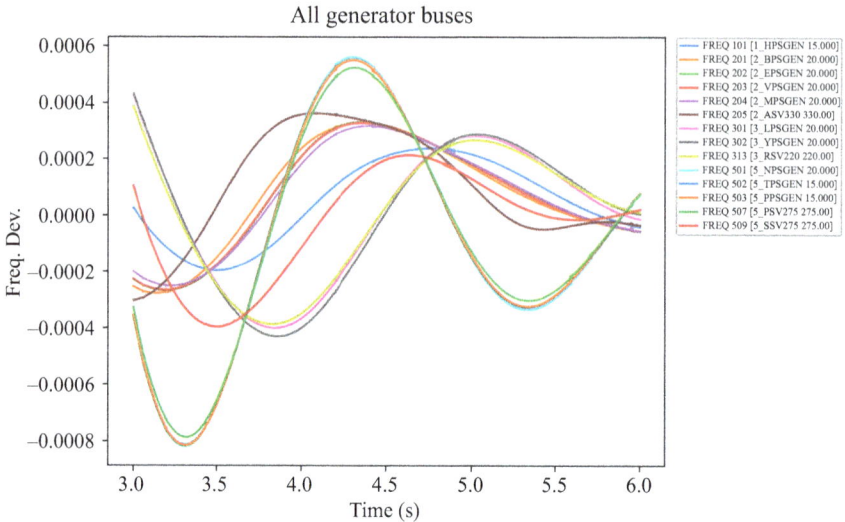

Figure 5.3   Frequency deviation measurements for all generator buses in the ES

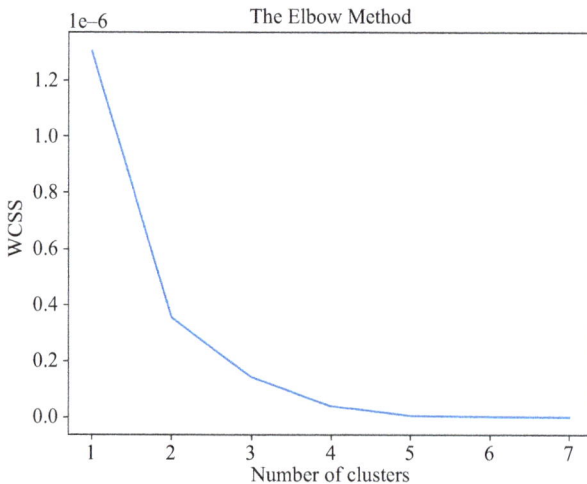

Figure 5.4   The elbow method for the identification of an optimum number of clusters

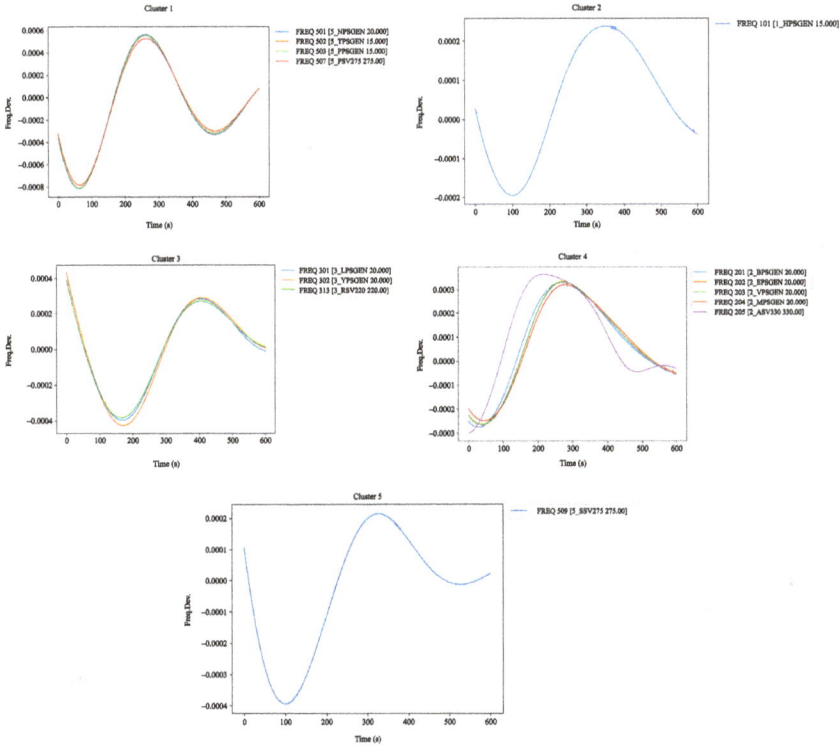

Figure 5.5   Coherent groups of generators in the ES

In Figure 5.4, the optimum number of clusters can be identified as 5. This is because after 5 number of clusters, the inertia of the dataset does not reduce; the elbow point is reached.

Figure 5.5 shows the clustered groups of generators in the ES using dynamic time warping specifying the number of clusters as 5.

From the results given in Figure 5.5, the coherent groups of generators in the ES are [501,502,503,507], [101], [301,302,313], [201,202,203,204,205], and [509]. For the reduction of the ES, these coherent groups could be aggregated and represented as equivalent generators. Therefore, 14 generators in the ES will be reduced to 5 equivalent generators. Zhukov's technique can be applied to reduce generator buses within a coherent group and Gaussian elimination technique can be used to eliminate the redundant (buses without current injections) in the ES.

5.5   Measurement-based reduction of wind power plants, solar power plants, microgrids, and ADNs

Since power systems are integrated with numerous large components with distinct dynamic characteristics, it is impractical to represent the entire ES with one equivalent model. The most appropriate way to reduce the ES is by reducing models with common dynamic characteristics together and represents them as equivalent models. Then these common dynamics can be aggregated to equivalent dynamics at the transmission level. Therefore, it is important to discuss other types of reduction strategies for unique entities of the power system such as wind power plants (WPPs), solar power plants (SPPs), microgrids (MGs), and ADNs. Measurement-based strategies are best suited for such entities because of the variability in operating points due to weather conditions.

5.5.1   *Wind farms*

Modern large-scale wind farms consist of thousands of wind turbines. Considering dynamic characteristics of individual wind turbine generator systems (WTGs) for dynamic analysis significantly increases the complexity of the analysis. However, the influence of wind farms in the dynamic analysis cannot be ignored. Thus, aggregated equivalent systems for WTGs are widely discussed. For the dynamic analysis, aggregated wind farm equivalent can be replaced at the point of common coupling.

Analysis of single-machine representation and multi-machine representation of large-scale doubly fed induction generator (DFIG) wind farms are given in [47]. The single-machine representation is useful when the wind farm has uniform wind velocity, whereas the multimachine representation is used for non-uniform wind farms. Multimachine representation is more accurate because uniform wind velocity is very rare in large-scale wind plants. In multi-machine representation, WTGs with the same operating points are lumped together to form an equivalent generator. Therefore, the unit division of WTGs in a large-scale wind farm is a crucial task in finding a multi-machine representation. For this, wind farms are analyzed in two arrangements known as "regular wind farms" and "irregular wind farms." Techniques used for unit division of WTGs in these two categories are discussed in [47].

In [48], a single-machine dynamic equivalence of a DFIG-based wind farm is obtained from a hybrid improved-particle swarm optimization genetic algorithm (HIGAPSO). This technique overcomes the shortcomings of two individual techniques genetic algorithm and particle swarm optimization. In this method, equivalent parameters are separately estimated for equivalent shaft, equivalent unit step-up transformer, and collector system.

In [49], a measurement-based dynamic equivalent is proposed by a deterministic autoregressive moving average model (DARMA) for a large-scale wind farm using the measurement taken at the point of common coupling (PCC). The cost function of the DARMA model is minimized using a recursive least squared method. The advantage of this technique is the dependency on measurement data only outside the wind farm, at the PCC. However, the dynamic equivalent model obtained from this analysis is a linear model, whereas the dynamics of a wind farm are inherently nonlinear.

5.5.2   *Solar farms*

The impact of solar farms in the dynamic stability analysis has become significant over the last few years due to the capacity of solar farms integrated into the grid. Similar to a WPP, an SPP can also be represented as a single-machine or a multi-machine system [50]. For a single-machine representation, there are two techniques, i.e., capacity equal weighted method (CEWM) and the parameter identification method. In the former, rated capacities of the inverters and transformers, equivalent capacitance, equivalent inductance, and equivalent impedance are calculated based on parameters of individual PV power units. In the latter, parameters for a single-machine equivalent system are estimated based on inputs to the PV plant and outputs from the PV plant.

Multi-machine representation is approved to be more accurate compared to single-machine representation. The first step in multi-machine representation is clustering PV groups. Several types of grouping indexes are proposed for clustering PV modules, i.e., feature distance based on control parameters, external characteristics, and iconic demarcation points. Out of these grouping indexes, the most widely used technique is the feature distance-based grouping. Once the grouping index is determined, the next step is clustering the PV modules based on a clustering algorithm. Clustering algorithms used in this context are K-means clustering, fuzzy clustering, spectral clustering, support vector machine clustering, canopy C-means, and affinity propagation [50,51]. Out of these techniques, the K-means clustering algorithm is the most widely used technique. Once clustering is completed, the next step is to find equivalent parameters. Equivalent parameters are determined like single-machine representation using a capacity equal-weighted method or parameter identification method.

5.5.3   *Microgrids*

A microgrid is an electrical system that contains loads, energy storage systems, and distributed energy resources (DERs) which operate in as a single controllable entity either being connected to the main grid or in an islanded model [52]. Due to the existence of DERs in a grid-connected microgrid, the interaction of its components with the grid must be considered in a stability analysis. Therefore, it is important to estimate an equivalent model for a grid-connected microgrid.

A black-box identification technique of the microgrid is proposed in [53] using Prony analysis and the least square optimization technique. The proposed technique can estimate the dynamics of the microgrid in terms of active power flow, reactive power flow, voltage, frequency, and current at the PCC, when the microgrid is subjected to internal disturbances. However, one of the drawbacks of Prony analysis is that it requires new Prony sets for each disturbance. Moreover, the technique proposed in [53] is only capable of predicting responses for small disturbances (e.g., load changes and changes to operating points of diesel generators) and cannot be used for electromechanical transient analysis of the main grid where large disturbances such as short circuit faults and tripping of generators are simulated.

An equivalent system for a grid-connected microgrid is derived from long short-term memory (LSTM) based neural networks in [54]. Datasets to train the LSTM are taken from the PCC where current at the PCC is the input to the LSTM and power flows (P, Q) are its outputs. Current inputs are categorized as real and imaginary components of current and previous inputs (which consist of four categories in total). Synthetic faults are generated in the distribution system at the PCC or in the microgrid itself to obtain the data for the training set. LSTM is a very powerful neural network technique that is famously used for sequence types of data, and its usage is validated with the accuracy of the predicted results. However, this type of equivalating of the microgrid is considered a black-box identification and hence lacks physical interpretability.

5.5.4 ADNs

ADNs are distribution networks that consist of distributed generation where bi-directional power flows can be observed between the transmission network and the distribution network [55]. The dynamic interaction between the distribution system and transmission system arises new problems to the dynamic state of the transmission system. In earlier studies, distribution networks were modeled as constant P and Q loads for dynamic simulations. However, the modeling of ADNs as such caused ramifications in the stability analysis.

An equivalent system based on Prony analysis and nonlinear least square optimization technique is explained in [56]. In this method, the Prony analysis is used to find initial estimates of the power flow of the ADNs. Unknown parameters such as amplitude, damping factor, frequency, and phase of the waveform are determined from the least square optimization algorithm using the measured data. In the parameter estimation process, parameters of individual ADN models are obtained for different operating conditions and averaged values of them are considered for the implementation of an equivalent ADN. It has been identified that by adjusting these averaged parameters, responses of the equivalent ADN for different operating scenarios can easily be obtained. However, the equivalent ADN obtained from this technique is only valid for small disturbances. Therefore, it is not useful for comprehensive stability studies of the transmission network such as system strength studies. Additionally, the parameter estimation technique is prone to convergence problems.

Although uncertainties of renewable energy sources due to weather forecasting errors in ADNs have been discussed in the literature, uncertainties occurring due to the spatial distance of renewable energy sources in ADNs are discussed in [57]. For the equivalent of the ADN, three sub-components are determined, i.e., equivalent wind turbine model, the equivalent PV system, and the equivalent ZIP model. Each sub-component consists of a deterministic part and an uncertain part. The parameters of the ADN are estimated online to adhere to time-varying loads and unintentional out of service of distributed generators. An enhanced reinforcement-based algorithm is incorporated for the parameter estimation technique.

An application of an RNN to represent the equivalent model of the ADNs is explained in [58]. In this application, the equivalent ADN model is represented as a differential-algebraic equation and is mapped to an RNN with three hidden layers. The input to the RNN is the voltage at the PCC and the output is either real power or reactive power at the PCC. However, RNNs are only capable of handling dynamic changes that are closer to the existing operating point. Therefore, this type of equivalent system can only be applied to voltage stability assessment or angle stability studies. To generalize the model for all applications, an Euclidean distance-based offline database is maintained. The Euclidean distance is calculated from measured active and reactive power flows at the PCC. If the distance between two operating points is larger than a threshold, a new RNN model is selected from the database relevant to that specific operating point. In this way, the RNN can accurately predict dynamics for a new set of disturbances in a different operating point apart from the current operation. Drawbacks of this technique are the extensive computational memory used to train RNNs and the lack of physical interpretability.

An LSTM-based RNN is adopted in finding the equivalent model of ADN in [59]. Due to the difficulties in traditional RNNs in identifying long-term dependencies, an LSTM block is used in this method to find the equivalent model. Like the method explained in [58], dynamics of the ADN are represented as a differential-algebraic equation and the LSTM block is mapped into those equations. Rather than selecting a fixed number of layers for the LSTM-based RNN, the number of layers is determined considering properties of the applied ADN such as scale, number of distributed generators, and number of dynamic loads or static loads. Input data for the RNN is normalized initially in the pre-processing step to fit with the characteristics of the activation function. Unlike the technique given in [58], the LSTM-based RNN can accurately predict dynamics of the ADN even for disturbances that were not used in the training process. However, the change of the operating point unfavorable for this method since the LSTM-based RNN model is valid for a fixed operating point.

An equivalent technique for the ADNs which focuses on changes to the operating point of the power system is proposed in [60]. For this, a repository of equivalent models for different operating points distinguished by a parameter vector is created. For a new operating condition, MC simulations are carried out for its set of training disturbances. From the results of MC simulations, two score functions are generated for active power and reactive power flows at the PCC. These score functions are evaluated for existing equivalent models in the repository for the same set of disturbances used in the new MC simulation. For an equivalent parameter set, if the worst score is bigger than the worst score of the new operating point, then the equivalent model with the highest score is selected from the repository. Otherwise, a new equivalent system is identified from a technique known as least absolute shrinkage and selection operator (LASSO). In the LASSO technique, the equivalent model with the highest score is selected as the reference model which is characterized by its parameter vector for the new operating point. Then the sensitivity of parameters in the reference parameter vector is evaluated to identify which parameters are most influential to the score function. Finally, only the significant parameters are updated to find the new equivalent model of the AND.

5.6   Case study: dynamic model reduction of the ES using LSTM recurrent neural networks

In this case study, the ES system of the AU14G system used for Section 5.3 is represented as an LSTM RNN model. This RNN is trained from three-phase to-ground faults simulated at all buses of the SS (there are 16 buses in the QLD system). All faults are similar in terms of time of the fault applied and the clearing time of the fault. A transmission line interconnecting the bus where the fault is applied is disconnected at the clearing time. Each dataset contains boundary measurements (V_s, δ_s, f_{dev}) and one tie-line power flows (P, Q) for 16 faults. The dataset is divided in half taking 8 faults for training and 8 faults for validation. The LSTM RNN model contains 1 input layer, 2 hidden recurrent layers of LSTM neurons, and 1 output layer. The input to the RNN taken as a 20 time-lagged (from t to $t + 20$) values of V_s, δ_s, f_{dev} and P. Each fault has recorded 1,205 measurement data during 6 seconds of the simulation. Therefore, the input dataset of the RNN $(N \times T \times D)$ is of shape $1,185 \times 20 \times 4$. The output of the RNN is used as the 21st value $(t + 21)$ of the P. Therefore, the output shape of the RNN has only 1 value.

After the training process is completed, a prediction was made for a fault at 401 as shown in Figure 5.6.

As seen from the figure, the LSTM-RNN can capture highly nonlinear dynamics of the ES with an accuracy of 0.89 R^2 value and 3.24 mean absolute error (MAE). Although the prediction is with a reasonable accuracy in terms of R^2 and MAE, the prediction during the transient period is very largely deviated. This could cause protection devices to activate in some case. Also, it should be noted that the

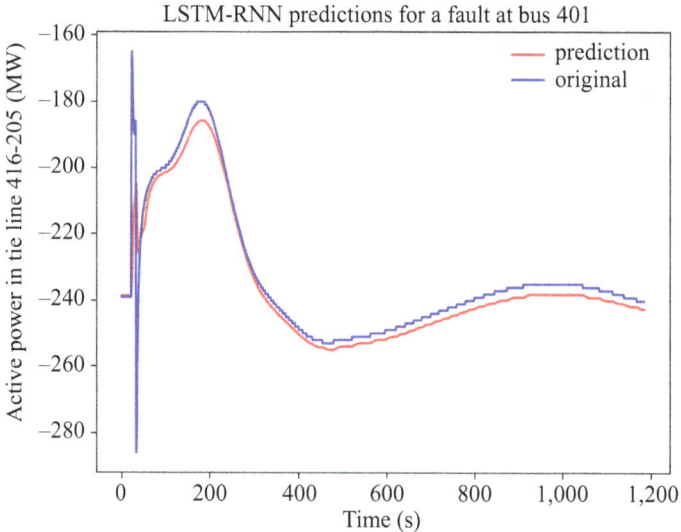

Figure 5.6   The prediction for active power flow using the LSTM-RNN for a fault at 401 of the SS

prediction was made for a fault that was used in the training process. Further investigations must be done for disturbances that were not used in the training phase as the accuracy usually depletes for dynamics that was not used in the training process.

5.7 Research gaps and conclusions

Dynamic model reduction of large power systems for SSA studies is a complex task due to various components integrated into the power systems such as wind farms, solar farms, microgrids, and ADNs that consist of various inverter-based generators (IBGs). Therefore, it is suggested in this chapter to identify the equivalent model of each of these components as seen from the transmission system bus before reducing the ES. As explained in the previous chapter, this problem cannot be tackled alone with classical techniques as they are incapable of adapting to operating conditions changes and lack of model data. In this chapter, the application of measurement-based techniques to solve the dynamic model reduction problem is comprehensively discussed that could also be applied to SSA. The application of measurement-based techniques involves two main streams in the dynamic model reduction, i.e., the black-box approach and the grey-box approach. In the black-box approach, novel machine learning techniques could be deployed to predict the dynamics of the external system (ES), which is the large part of the power network external to the part of the system under study. The parameter estimation technique is applicable in both black-box and grey-box situations to estimate the parameters of an equivalent model, to find the poles and zeros of a transfer function model or find parameters of a differential algebraic model. In this regard, novel biologically inspired metaheuristic techniques that can converge faster and approach the global minimum than traditional nonlinear least square technique. Although, the measurement-based technique is promising in for capturing dynamics of an ES, they inhibit following major drawbacks:

- Machine learning models must be extensively trained that could take long times.
- Machine learning models deplete their accuracy for predictions that are not used in the training process. However, all possible disturbances in a power system cannot be used for training machine learning model as there can be many disturbance scenarios in a power system.
- Machine learning models do not have a physical interpretability, i.e., they cannot be referred to actual components of a power system.
- Parameter estimation can lead to local optimum in some cases even though how sophisticated the parameter estimation technique is.
- Parameter estimation can take a long time if the objective vector of the technique consists of many parameters. For example, if a synchronous generator model is used to estimate the parameters, there are several parameters associated with its subcomponents such as turbine governor, exciter, power system stabilizer, and rotor.

Considering these drawbacks, it could be suggested that the best way of reducing an ES model is a combination of classical and data-driven techniques. This is because in most cases, some data of the ES is available for the model reduction. Only a few dynamic data are unknown such as few generator dynamics that can be approximated with data-driven techniques. In the parameter estimation process, a set of critical parameters should be selected based on sensitivity analysis to decrease the burden as well as reaching the local optimum. In this way, a robust model of the ES can be implemented for SSA studies.

References

[1]   A. Géron, *Hands-On Machine Learning with Scikit-Learn, Keras, and TensorFlow*. O'Reilly Media, Inc., 2022.

[2]   S. V. Mahadevkar, B. Khemani, S. Patil, *et al.*, "A review on machine learning styles in computer vision-techniques and future directions," *IEEE Access*, vol. 10, pp. 107293–107329, 2022.

[3]   K. Amarasinghe, D. L. Marino, and M. Manic, "Deep neural networks for energy load forecasting," in *2017 IEEE 26th International Symposium on Industrial Electronics (ISIE)*, 2017. New York, NY: IEEE, pp. 1483–1488.

[4]   D. Díaz-Vico, A. Torres-Barrán, A. Omari, and J. R. Dorronsoro, "Deep neural networks for wind and solar energy prediction," *Neural Processing Letters*, vol. 46, pp. 829–844, 2017.

[5]   Z. Zheng, Y. Yang, X. Niu, H.-N. Dai, and Y. Zhou, "Wide and deep convolutional neural networks for electricity-theft detection to secure smart grids," *IEEE Transactions on Industrial Informatics*, vol. 14, no. 4, pp. 1606–1615, 2017.

[6]   Z. Shi, W. Yao, L. Zeng, *et al.*, "Convolutional neural network-based power system transient stability assessment and instability mode prediction," *Applied Energy*, vol. 263, p. 114586, 2020.

[7]   O. Malik, "An adaptive power system stabilizer based on recurrent neural networks," *IEEE Transactions on Energy Conversion*, vol. 12, no. 4, pp. 413–418, 1997.

[8]   M. N. Fekri, H. Patel, K. Grolinger, and V. Sharma, "Deep learning for load forecasting with smart meter data: online adaptive recurrent neural network," *Applied Energy*, vol. 282, p. 116177, 2021.

[9]   Y. Wang, R. Zou, F. Liu, L. Zhang, and Q. Liu, "A review of wind speed and wind power forecasting with deep neural networks," *Applied Energy*, vol. 304, p. 117766, 2021.

[10] Y. Jung, J. Jung, B. Kim, and S. Han, "Long short-term memory recurrent neural network for modeling temporal patterns in long-term power forecasting for solar PV facilities: case study of South Korea," *Journal of Cleaner Production*, vol. 250, p. 119476, 2020.

[11] S. Wen, Y. Wang, Y. Tang, Y. Xu, P. Li, and T. Zhao, "Real-time identification of power fluctuations based on LSTM recurrent neural network: a

case study on Singapore power system," *IEEE Transactions on Industrial Informatics*, vol. 15, no. 9, pp. 5266–5275, 2019.

[12] V. Veerasamy, N. I. A. Wahab, M. L. Othman, *et al.*, "LSTM recurrent neural network classifier for high impedance fault detection in solar PV integrated power system," *IEEE Access*, vol. 9, pp. 32672–32687, 2021.

[13] A. Ayad, H. E. Farag, A. Youssef, and E. F. El-Saadany, "Detection of false data injection attacks in smart grids using recurrent neural networks," in *2018 IEEE Power & Energy Society Innovative Smart Grid Technologies Conference (ISGT)*, 2018. New York, NY: IEEE, pp. 1–5.

[14] Z. Sun, S. Zhao, and J. Zhang, "Short-term wind power forecasting on multiple scales using VMD decomposition, K-means clustering and LSTM principal computing," *IEEE Access*, vol. 7, pp. 166917–166929, 2019.

[15] Y. Wang, D. Wang, and Y. Tang, "Clustered hybrid wind power prediction model based on ARMA, PSO-SVM, and clustering methods," *IEEE Access*, vol. 8, pp. 17071–17079, 2020.

[16] D. Liu and K. Sun, "Random forest solar power forecast based on classification optimization," *Energy*, vol. 187, p. 115940, 2019.

[17] S. M. Zali and J. V. Milanović, "Generic model of active distribution network for large power system stability studies," *IEEE Transactions on Power Systems*, vol. 28, no. 3, pp. 3126–3133, 2013.

[18] A. M. Stankovic, A. T. Saric, and M. Milosevic, "Identification of nonparametric dynamic power system equivalents with artificial neural networks," *IEEE Transactions on Power Systems*, vol. 18, no. 4, pp. 1478–1486, 2003.

[19] A. M. Saxe, Y. Bansal, J. Dapello, *et al.*, "On the information bottleneck theory of deep learning," *Journal of Statistical Mechanics: Theory and Experiment*, vol. 2019, no. 12, p. 124020, 2019.

[20] N. Tong, Z. Jiang, S. You, *et al.*, "Dynamic equivalence of large-scale power systems based on boundary measurements," in *2020 American Control Conference (ACC)*, 2020. New York, NY: IEEE, pp. 3164–3169.

[21] J. Kim and N. Kasabov, "HyFIS: adaptive neuro-fuzzy inference systems and their application to nonlinear dynamical systems," *Neural Networks*, vol. 12, no. 9, pp. 1301–1319, 1999.

[22] H. Shakouri and H. R. Radmanesh, "Identification of a continuous time nonlinear state space model for the external power system dynamic equivalent by neural networks," *International Journal of Electrical Power & Energy Systems*, vol. 31, nos. 7–8, pp. 334–344, 2009.

[23] D. Kler, Y. Goswami, K. Rana, and V. Kumar, "A novel approach to parameter estimation of photovoltaic systems using hybridized optimizer," *Energy Conversion and Management*, vol. 187, pp. 486–511, 2019.

[24] T. J. Ypma, "Historical development of the Newton–Raphson method," *SIAM Review*, vol. 37, no. 4, pp. 531–551, 1995.

[25] K. Et-Torabi, I. Nassar-Eddine, A. Obbadi, *et al.*, "Parameters estimation of the single and double diode photovoltaic models using a Gauss–Seidel

algorithm and analytical method: a comparative study," *Energy Conversion and Management*, vol. 148, pp. 1041–1054, 2017.

[26] S. Mirjalili, "Genetic algorithm," in *Evolutionary Algorithms and Neural Networks*. New York, NY: Springer, 2019, pp. 43–55.

[27] J. C. Bansal, "Particle swarm optimization," in *Evolutionary and Swarm Intelligence Algorithms*. New York, NY: Springer, 2019, pp. 11–23.

[28] K. Hussain, M. N. Mohd Salleh, S. Cheng, and Y. Shi, "Metaheuristic research: a comprehensive survey," *Artificial Intelligence Review*, vol. 52, no. 4, pp. 2191–2233, 2019.

[29] B. Morales-Castañeda, D. Zaldivar, E. Cuevas, F. Fausto, and A. Rodríguez, "A better balance in metaheuristic algorithms: Does it exist?," *Swarm and Evolutionary Computation*, vol. 54, p. 100671, 2020.

[30] A. Biswas, K. Mishra, S. Tiwari, and A. Misra, "Physics-inspired optimization algorithms: a survey," *Journal of Optimization*, vol. 2013, 2013.

[31] A. Rahim and A. Al-Ramadhan, "Dynamic equivalent of external power system and its parameter estimation through artificial neural networks," *International Journal of Electrical Power & Energy Systems*, vol. 24, no. 2, pp. 113–120, 2002/2/1.

[32] M. J. Orr, "Introduction to radial basis function networks," Technical Report, Center for Cognitive Science, University of Edinburgh, 1996.

[33] M. Shiroei, B. Mohammadi-Ivatloo, and M. Parniani, "Low-order dynamic equivalent estimation of power systems using data of phasor measurement units," *International Journal of Electrical Power & Energy Systems*, vol. 74, pp. 134–141, 2016.

[34] A. T. Sarić, M. T. Transtrum, and A. M. Stanković, "Data-driven dynamic equivalents for power system areas from boundary measurements," *IEEE Transactions on Power Systems*, vol. 34, no. 1, pp. 360–370, 2018.

[35] J. J. Moré, "The Levenberg-Marquardt algorithm: implementation and theory," in *Numerical Analysis*. New York, NY: Springer, 1978, pp. 105–116.

[36] X. Zhang, Y. Xue, S. You, *et al.*, "Measurement-based power system dynamic model reductions," in *2017 North American Power Symposium (NAPS)*, 2017.

[37] Z. Jiang, N. Tong, Y. Liu, Y. Xue, and A. G. Tarditi, "Enhanced dynamic equivalent identification method of large-scale power systems using multiple events," *Electric Power Systems Research*, vol. 189, p. 106569, 2020.

[38] L. Aththanayake, A. Mahmud, N. Hosseinzadeh, and A. Gargoom, "Performance analysis of regression and artificial neural network schemes for dynamic model reduction of power systems," in *2021 3rd International Conference on Smart Power & Internet Energy Systems (SPIES)*, 2021. New York, NY: IEEE, pp. 358–363.

[39] A. M. Khalil and R. Iravani, "Power system coherency identification under high depth of penetration of wind power," *IEEE Transactions on Power Systems*, vol. 33, no. 5, pp. 5401–5409, 2018.

[40] S. Aghabozorgi, A. S. Shirkhorshidi, and T. Y. Wah, "Time-series clustering–a decade review," *Information Systems*, vol. 53, pp. 16–38, 2015.

[41] C. A. Ratanamahatana and E. Keogh, "Making time-series classification more accurate using learned constraints," in *Proceedings of the 2004 SIAM International Conference on Data Mining*, 2004. Philadelphia, PA: SIAM, pp. 11–22.

[42] H. Akoglu, "User's guide to correlation coefficients," *Turkish Journal of Emergency Medicine*, vol. 18, no. 3, pp. 91–93, 2018.

[43] S. Hwang, "Time series models for forecasting construction costs using time series indexes," *Journal of Construction Engineering and Management*, vol. 137, no. 9, pp. 656–662, 2011.

[44] M. R. Berthold and F. Höppner, "On clustering time series using Euclidean distance and Pearson correlation," 2016, arXiv preprint arXiv:1601.02213.

[45] M. Richardson, "Principal component analysis," URL: http://people. maths. ox. ac. uk/richardsonm/SignalProcPCA. pdf (last access: 3.5.2013). Aleš Hladnik Dr., Asst. Prof., Chair of Information and Graphic Arts Technology, Faculty of Natural Sciences and Engineering, University of Ljubljana, Slovenia ales. hladnik@ ntf. uni-lj. si, vol. 6, p. 16, 2009.

[46] M. Gibbard and D. Vowles, "IEEE PES task force on benchmark systems for stability controls simplified 14-generator model of the south east Australian power system," in *The University of Adelaide*, vol. 5005, 2014.

[47] J. Zou, C. Peng, Y. Yan, H. Zheng, and Y. Li, "A survey of dynamic equivalent modeling for wind farm," *Renewable and Sustainable Energy Reviews*, vol. 40, pp. 956–963, 2014.

[48] Y. Zhou, L. Zhao, and W.-J. Lee, "Robustness analysis of dynamic equivalent model of DFIG wind farm for stability study," *IEEE Transactions on Industry Applications*, vol. 54, no. 6, pp. 5682–5690, 2018.

[49] D.-E. Kim and M. A. El-Sharkawi, "Dynamic equivalent model of wind power plant using parameter identification," *IEEE Transactions on Energy Conversion*, vol. 31, no. 1, pp. 37–45, 2015.

[50] P. Han, Z. Lin, L. Wang, G. Fan, and X. Zhang, "A survey on equivalence modeling for large-scale photovoltaic power plants," *Energies*, vol. 11, no. 6, p. 1463, 2018.

[51] P. Chao, W. Li, X. Liang, Y. Shuai, F. Sun, and Y. Ge, "A comprehensive review on dynamic equivalent modeling of large photovoltaic power plants," *Solar Energy*, vol. 210, pp. 87–100, 2020.

[52] M. Farrokhabadi, C. A. Canizares, J. W. Simpson-Porco, *et al.*, "Microgrid stability definitions, analysis, and examples," *IEEE Transactions on Power Systems*, vol. 35, no. 1, pp. 13–29, 2019.

[53] P. N. Papadopoulos, T. A. Papadopoulos, P. Crolla, A. J. Roscoe, G. K. Papagiannis, and G. M. Burt, "Black-box dynamic equivalent model for microgrids using measurement data," *IET Generation, Transmission & Distribution*, vol. 8, no. 5, pp. 851–861, 2014.

[54] C. Cai, H. Liu, Y. Tao, Z. Deng, W. Dai, and J. Chen, "Microgrid equivalent modeling based on long short-term memory neural network," *IEEE Access*, vol. 8, pp. 23120–23133, 2020.

[55] S. Chowdhury, S. Chowdhury, and P. Crossley, *Microgrids and Active Distribution Networks*, 2022. London: IET.

[56] S. M. Zali and J. Milanovic, "Dynamic equivalent model of distribution network cell using Prony analysis and nonlinear least square optimization," in *2009 IEEE Bucharest PowerTech*, 2009: IEEE, pp. 1–6.

[57] X. Shang, Z. Li, J. Zheng, and Q. Wu, "Equivalent modeling of active distribution network considering the spatial uncertainty of renewable energy resources," *International Journal of Electrical Power & Energy Systems*, vol. 112, pp. 83–91, 2019.

[58] C. Zheng, S. Wang, Y. Liu, and C. Liu, "A novel RNN based load modelling method with measurement data in active distribution system," *Electric Power Systems Research*, vol. 166, pp. 112–124, 2019.

[59] C. Zheng, S. Wang, Y. Liu, *et al.*, "A novel equivalent model of active distribution networks based on LSTM," *IEEE Transactions on Neural Networks and Learning Systems*, vol. 30, no. 9, pp. 2611–2624, 2019.

[60] G. Chaspierre, G. Denis, P. Panciatici, and T. Van Cutsem, "A dynamic equivalent of active distribution network: derivation, update, validation and use cases," *IEEE Open Access Journal of Power and Energy*, vol. 8, pp. 497–509, 2021.

Chapter 6

Inverter-based resources and their impact on power system inertia and system strength

Ameen Gargoom[1], Nasser Hosseinzadeh[1], Hassan Haes Alhelou[2], Behrooz Bahrani[2] and Ehsan Farahani[3]

Abstract

This chapter provides an overview of the impact of integrating inverter-based resources (IBRs) on power system inertia and strength. Inertia is a crucial aspect in power systems for maintaining frequency stability, and the chapter explores the historical perspectives and importance of system inertia. In addition, with the increasing integration of IBRs, the chapter examines how they impact the power system's inertia and strength, highlighting the issues arising from the low inertia provided by IBRs and the reduction in the fault current contribution. Furthermore, the chapter delves into the frequency response and inertia requirements and presents different methods for estimating power system inertia. The case study of a power system with integrated wind energy plants illustrates the impact of IBRs on the power system's stability. Finally, the chapter discusses research gaps and industry challenges, as well as future research directions. The presented information provides valuable insights for power system operators and researchers to better manage the challenges posed by integrating IBRs into power systems.

6.1 Introduction

6.1.1 What is inertia, and why is it important in the power system?

The power system industry is undergoing a rapid transformation, shifting it into a new era of operation and analysis due to the increased penetrations of the distribution of energy resources (DERs) and inverter-based resources (IBRs) at all levels. This shift is introducing new challenges which were not significant in traditional power systems. One of these challenges is the reduction in power system

[1]School of Engineering, Deakin University, Australia
[2]School of Electrical and Computer System Engineering, Monash University, Australia
[3]Australian Energy market Operator (AEMO)

inertia which refers to the inherent ability of a power system to resist changes in frequency when there is a disturbance in the system, such as the sudden loss or addition of generation or load. It is a result of the physical properties of the rotating masses in generators, which serve to store kinetic energy and provide a buffer against frequency fluctuations. Simultaneously, IBRs contribute to the reduction of power system strength, which causes voltage fluctuations when the system undergoes a disturbance. This chapter focuses on the inertia problem.

In power systems, the amount of inertia present is a critical factor in maintaining system stability and ability to maintain a consistent frequency. When there is a disturbance in the power system, such as a fault or sudden change in loads, the frequency of the system will change. The amount of inertia present in the system determines how much the frequency changes and how fast this frequency will change in response to the disturbance. A system with a large amount of inertia will respond more slowly to frequency changes and experience smaller frequency deviations, while a system with low inertia will experience more rapid and larger frequency deviations.

Maintaining a stable frequency is critical to the operation of the power system. If the frequency deviates too much from its nominal value (typically 50 or 60 Hz), it can cause significant issues, including tripping of protection devices, damage to equipment, and even blackouts. Therefore, the amount of inertia in the system is a key factor in ensuring the stability and reliability of the power system.

There have been several events in power systems in which the low inertia played a significant role in the severity and duration of the event. Below are some of these events.

1. The 2003 blackout in the UK [1]: This event was caused by the tripping of two high-voltage transmission lines. The low inertia of the system contributed to the rapid frequency drop and the subsequent loss of generation, leading to an automatic load shedding that affected around 476,000 customers.
2. The 2011 blackout in Southern California [2]: This event was caused by a series of equipment failures that triggered a chain reaction of power outages. The low inertia of the system allowed the frequency to drop quickly, leading to an automatic load shedding and a widespread blackout.
3. The South Australian blackout in 2016 [3]: This event was caused by a severe storm that damaged transmission lines and caused multiple generators to trip offline. The low inertia of the system exacerbated the frequency drop, leading to automatic load shedding and a total blackout.
4. The 2018 power outage in Puerto Rico [4]: This event was triggered by damage to the transmission and distribution system from Hurricane Maria. The low inertia of the system made it difficult to re-start the generators, resulting in an extended outage that lasted for months.
5. The Texas power crisis in 2021 [5]: The low inertia of the Texas power system contributed to the failure of multiple generators during a severe winter storm, leading to widespread outages and a near-total collapse of the system.

In addition, reduced system strength can also be considered as an inherited issue from replacing a large number of synchronous generators with IBRs and the consequent reduction in system inertia. The reduction in system inertia can lead to a decrease in the system's ability to withstand and recover from disturbances, making power systems more vulnerable to frequency and voltage instability. This is because inertia plays a crucial role in maintaining the stability of the grid's frequency and voltage levels, which are critical for maintaining the quality and reliability of electricity supply.

6.1.2   Historical perspectives

Traditionally, inertia was provided in the power system by synchronous generators which are large rotating machines that are directly coupled to a turbine and rotate at the same speed as the power system. They provide not only power but also a significant amount of inertia to the system. The amount of inertia provided by a synchronous generator is proportional to its size and rotational speed. Therefore, large synchronous generators with high rotational speeds provide the most inertia.

In the past, the power system relied heavily on synchronous generators for both power and inertia. The number of synchronous generators connected to the power system was high, and their large rotational inertia helped to maintain system stability in the face of disturbances. However, in recent years, there has been a shift towards using asynchronous generation sources such as wind and solar power. These types of generation sources are usually not directly connected to the grid and do not provide the same level of inertia as synchronous generators. This has resulted in a decrease in the overall amount of inertia in the power system.

The decrease in the amount of synchronous generation and inertia in the power system have created new challenges for maintaining system stability. The issue of decreased inertia has been ranked the top issue in modern power systems by European transmission system operators (TSOs) as published by the European Network of Transmission System Operators (ENTSO-E) as listed in Table 6.1 [6]. Therefore, as the power system becomes more reliant on IBRs (asynchronous generation), new techniques, and technologies are needed to ensure that system stability is maintained. These include the use of energy storage systems, advanced control strategies, and the development of new types of generation that can provide more inertia to the system.

Table 6.1   Top issues impacting power system stability [6]

Rank	Score	Issue
1	17.35	Decrease of inertia
2	10.16	Resonance caused by power electronics and cables
3	9.84	Reduced transient stability margins
4	8.91	Power electronics-connected generators' participation in frequency containment is either absent or incorrect
5	8.19	Interaction between power electronics controllers and passive AC components

6.1.3   How IBRs impact power system inertia?

IBRs have a significant impact on power system inertia. Unlike traditional synchronous generators, which have a large rotating mass that provides inertia to the power system, IBRs do not have the same level of inertial response and thus their response to changes in frequency is typically faster. This is because IBRs are connected to the grid through power electronics, such as inverters, which do not have the same level of mechanical inertia as a synchronous generator. This means that if there is a sudden change in load or a fault in the system, with a proper control system, IBRs can respond quickly and reduce the frequency deviation. However, if there is a significant loss of generation on the system, the reduction in inertia from the loss of synchronous generators can lead to a rapid drop in frequency.

As more IBRs, such as solar PV and wind power, are connected to the grid, the overall inertia of the system decreases. This can have significant impacts on the stability and security of the power system, as lower levels of inertia increase the susceptibility of the power system to rapid changes in frequency (RoCoF) due to imbalances in supply and demand. This can lead to cascading failures and blackouts.

To mitigate the impact of IBRs on power system inertia, new technologies and strategies are being developed. One approach is to use advanced power electronics and control systems to provide synthetic or virtual inertia. This involves mimicking the inertial response of synchronous generators using power electronics and control systems. Another approach is to combine IBRs with energy storage systems, such as batteries, which can provide rapid response to frequency deviations and help to stabilize the power system. This chapter will explore some of these technologies.

6.1.4   How IBRs impact power system strength?

In a simple term, power system strength refers to the ability of a system to maintain its voltage level during the dynamics of power systems. A system that is sensitive to voltage changes is considered weak and vulnerable to instability, while a system that is less susceptible to voltage variations is referred to as a strong system. Keeping a high level of system strength is important for power system operation to maintain voltage stability, especially in modern power systems with high levels of IBRs and DERs.

Traditionally, system strength has been assessed at nodes using the fault level at that node, with large fault currents indicating a strong system and small fault currents indicating a weak system. Historically, synchronous generators have been major contributors to fault currents and, thus, system strength. However, the large-scale integration of IBRs has impacted system strength in two significant ways.

First, IBRs typically have low or zero inertia, which can reduce the system's ability to withstand and recover from disturbances, making it more susceptible to frequency and voltage instability. The reduction in system inertia can result in decreased system stability and increased vulnerability to disturbances. Second, IBRs do not typically contribute large fault currents due to the limitations of power electronics, which can significantly reduce the strength of the system if the penetration of IBRs is high. This can result in a decrease in fault levels and a potential loss of system strength.

Overall, the impacts of IBRs on power system inertia and strength are important areas of research that require further investigation and development of effective mitigation strategies.

6.2   Frequency response and inertia

Frequency stability refers to the ability of a power system to maintain a stable frequency in the event of disturbances, such as sudden changes in load or generation. The initial behavior of the frequency following a disturbance is determined by the power system's inertia. Thus, frequency response is closely related to the power system inertia. In general, frequency response is determined based on the automatic adjustment of the generation-load balance in response to changes in frequency alone, which is measured in megawatts (MW) [7]. For power system operation, frequency response is a critical aspect of the stability and reliability. If a disturbance causes a power deficit, the power system's inherent kinetic energy is released from the rotating masses of the synchronous machines to the grid. This causes the synchronous machines to slow down (governed by the inertia), leading to a decline in the system frequency, also known as under-frequency. On the other hand, if there is a power surplus, the power system's inherent kinetic energy is absorbed by the rotating masses of the synchronous machines, causing their speed to increase, leading to an increase in system frequency, also known as over-frequency.

In an interconnected power system, the frequency of the entire system is dependent on the balance between supply and demand. If there is a sudden imbalance, the frequency will deviate from its nominal value. This deviation can cause problems for both the power system and the equipment connected to it. Frequency response, therefore, is important because it helps maintain a stable frequency. The faster a power system can respond to a disturbance, the less the frequency will deviate from its nominal value. This helps prevent equipment from tripping offline or malfunctioning, which can cause blackouts or damage to the equipment.

To achieve proper frequency response, power systems rely on a combination of primary, secondary, and tertiary control. Primary control refers to the inherent response of generators to changes in frequency due to their inherent rotational inertia. Secondary control involves the use of automatic generation control (AGC) to adjust generation output to maintain a stable frequency. Tertiary control involves the use of reserves, such as frequency containment reserves (FCR) and frequency restoration reserves (FRR), to provide additional power during periods of high demand or low generation.

Another important characteristic of frequency response is the rate of change of frequency (RoCoF). RoCoF is a measure of how quickly the frequency changes in the event of a disturbance. It is derivative of the frequency with respect to time, and it gives an indication of how fast the frequency is moving away from the nominal value. A high RoCoF can cause problems for sensitive equipment, such as variable

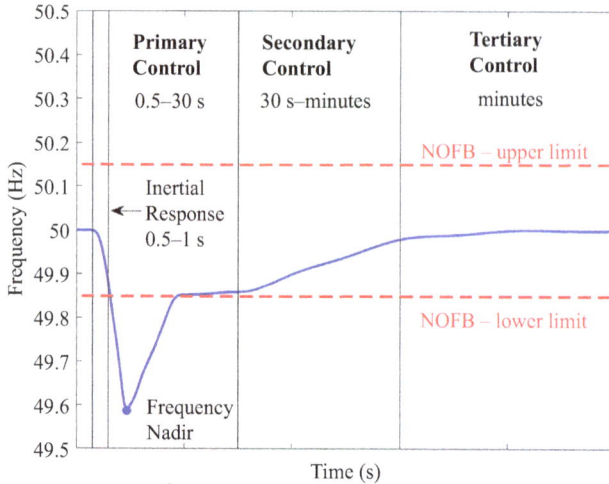

Figure 6.1　Frequency response in power systems [8]

speed drives, and can lead to a loss of power system stability. In the event of a sudden loss of generation, the RoCoF increases rapidly, and this can result in instability and potential cascading outages. Therefore, it is important for the power system to have sufficient resources to provide fast frequency response to arrest the RoCoF and stabilize the system.

Power system operators often set limits on the maximum allowable RoCoF to ensure the system remains stable and reliable. The exact value of the RoCoF limit varies depending on the specific power system, but, in general, faster RoCoFs require more rapid and powerful frequency response resources to maintain system stability.

The other important aspect of frequency response is frequency nadir which refers to the lowest frequency reached during a disturbance. The frequency nadir is an important parameter for assessing the stability and resilience of the power system. If the frequency nadir is too low, it can cause equipment to trip offline, leading to further disturbances and blackouts. Figure 6.1 shows the timescale of the three control stages of a typical frequency response for a 50 Hz system.

Therefore, understanding the level of power system inertia can help power system operators predict how much and how quickly the frequency will drop during the initial moments after a disturbance. With this information, operators can take appropriate actions to address the issue.

6.3　Inertia requirement

In the past, it was not necessary to require power system participants to provide inertia to the power grid because there were many synchronous generators connected to the grid that could provide the required inertia for power system security.

However, the large integration of IBRs and their impact on the system inertia (and thus stability) has introduced new requirements for power system inertia. Requirements for power systems inertia vary by country, as different countries have different energy mix and grid configurations. However, there are some commonalities in the approaches taken by different countries.

In the United States, the North American Electric Reliability Corporation (NERC) has introduced standards for maintaining system inertia. The standard requires utilities to maintain a minimum amount of inertia for the system to remain stable during contingencies. The amount of required inertia varies by region and system configuration. Some of these standards are:

- BAL-003-1: Frequency Response and Frequency Bias Setting (2015): This standard sets requirements for the frequency response of generators and other resources in the system. It also requires balancing authorities to set appropriate frequency bias settings for generators, which ensures that the generators respond appropriately to changes in frequency.
- BAL-002-WECC-3: Contingency Reserve (2019): This standard requires balancing authorities to maintain sufficient contingency reserves to respond to sudden system disturbances that affect the frequency, such as the loss of a large generator. The standard also requires balancing authorities to ensure that these reserves are sufficient to meet the minimum inertia requirements for the system.

In the United Kingdom, National Grid Electricity System Operator (ESO) has introduced a minimum inertia requirement for the grid. The requirement states that the system should have at least 3.6 GW of inertia available at any time to maintain frequency stability. The 3.6 GW-seconds requirement applies to the entire electricity transmission system in the United Kingdom, which includes both the high-voltage transmission network and the interconnectors that link the United Kingdom to other countries. The requirement is periodically reviewed by National Grid ESO to ensure that it remains appropriate for the evolving needs of the grid [9].

In Australia, the National Electricity Market (NEM) is the wholesale electricity market that operates in the eastern and south-eastern states of Australia. The NEM has established a minimum threshold level of inertia required to operate an islanded inertia sub-network in a satisfactory operating state. This minimum inertia level is defined in Clause 5.20B.2(b)(1) of the National Electricity Rules (NER). The NER specifies that the minimum inertia level must be maintained to ensure that the system can withstand a contingency event, such as the loss of a large generator or transmission line. The Australian Energy Market Operator (AEMO) is responsible for the overall management of the NEM, including ensuring the stability and reliability of the power system. The AEMO has introduced a minimum inertia requirement for the NEM to maintain frequency stability. The minimum inertia requirement is calculated based on a range of factors, including the level of intermittent generation and the size and configuration of the power system. According to the Inertia Requirements Methodology document produced by AEMO, the minimum inertia constant (MIC) for the NEM is currently set at 6 GW-seconds.

However, the actual required level of inertia for each region within the NEM can vary depending on the system characteristics and contingency scenarios.

To ensure that there is sufficient system inertia, the AEMO regularly monitors the available inertia levels and identifies any potential shortfall in different parts of the NEM. The AEMO has identified several regions where there is an inertia shortfall and has implemented measures to increase the level of system inertia in those regions. For example, in the South Australian region, the AEMO has implemented additional inertia support activities to address the potential shortfall.

Inertia support activities in the NEM include Contingency Frequency Control Ancillary Services (FCAS) and Fast Frequency Response (FFR). Contingency FCAS is a market mechanism that provides automatic generation control services to respond to frequency deviations caused by contingencies. FFR is a market mechanism that provides rapid response services to changes in system frequency. These services can help to provide additional inertia support to the system during contingency events and reduce the risk of instability.

The level of system inertia is related to the fast frequency control ancillary service (FCAS) requirement. The Fast FCAS specification in the NEM mandates complete delivery within six seconds. The previous approach for calculating inertia requirements assumes that the full six seconds is required for a complete Fast FCAS response. However, if the FCAS response can be delivered fully in less than 6 s, this can result in meeting the Frequency Standards with dispatch of less inertia, for a given contingency size.

The minimum threshold level of inertia for each state in Australia can be found in the AEMO's Inertia Requirements Methodology document is given in Table 6.2 [10]. For example, in the South Australian region, the available inertia through system strength (*which is the amount of inertia available through system strength is the amount of inertia that can be provided by the synchronous generators in the system without any additional support or equipment*) is around 4,900 MWs, while the minimum threshold level of inertia is around 4,400 MWs. This means that the available inertia is low and close to the minimum threshold level, which makes the system frequency stability close to a critical state. Note that those figures of the available inertia are updated regularly by AEMO based on the network condition.

Table 6.2   Inertia requirements for Australia based on the 2018 AEMO's report [10]

Inertia sub-networks (regions)	Inertia available through system strength (MWs)	Minimum threshold level of inertia (MWs)	Secure operating level of inertia (MWs)
Queensland	11,950	12,800	16,000
New South Wales	18,100	10,000	12,500
Victoria	10,900	12,600	15,400
South Australia	4,900	4,400	6,000
Tasmania	2,000	3,200	3,800

6.4 Estimation methods of power system inertia

Inertia estimation is essential for grid operators to ensure grid stability and maintain reliable power supply. There are several methods for estimating system inertia, which can be broadly classified into two categories: model-based and measurement-based methods [11]. Model-based methods rely on mathematical models of power system components and their behavior to estimate system inertia. These models typically consider the physical properties of the components, such as their mass, moment of inertia, and damping characteristics, and can be used to estimate the inertia of individual components, such as generators, as well as the overall system inertia. Accurate estimation of inertia using model-based methods is important for effective control of frequency and power flow in the system.

Measurement-based methods, on the other hand, rely on direct measurements of system variables to estimate inertia. Those methods estimate the inertia of a power system through direct measurements of system variables. They can estimate both individual component and system inertia and are often combined with model-based methods for higher accuracy. Wide-area monitoring systems, which use phasor measurement units (PMUs), are becoming more prevalent in several countries. PMUs can provide information that can be used to estimate the inertial constants of large power systems quickly and reliably [12].

Disturbance-based estimation is a widely used measurement-based method for estimating power system inertia. The method involves measuring the frequency response of the power system following a disturbance [13]. Specifically, the RoCoF and the initial amount of power imbalance caused by the disturbance are measured. To ensure accurate estimation, measurements are taken promptly after the disturbance to avoid the effects of frequency response and load changes. By analyzing the relationship between the disturbance and the system response, inertia can be estimated using various algorithms and techniques. Disturbance-based estimation is often preferred due to its simplicity, low cost, and ability to estimate inertia in real-time.

However, the method has some limitations and challenges, including the need for precise measurements and the impact of system uncertainties and nonlinearities. It can be challenging to identify the exact moment of a disturbance, which is necessary for an accurate estimation. Another challenge is the poor quality of frequency measurements, which can include transients, oscillatory components, and noise. This can lead to inaccuracies in the estimation of RoCoF, which is critical for an accurate estimation of inertia. Different strategies, such as curve fitting or low-pass filters, can be employed to eliminate these fluctuations and improve the accuracy of the measurement.

Additionally, the accurate amount of power imbalance and corresponding RoCoF value at the same time should be available to estimate power system inertia. However, the power imbalance can deviate significantly from the disturbance value due to the response of generators' governors and the frequency and voltage dependencies of the loads. This issue has not yet been fully addressed in the literature. Moreover, the frequency deviation will have different values at different parts of the grid for a certain time, so different frequency measurements from various parts of the grid are required to accurately represent the system's average frequency. The Centre of Inertia (COI) can be used to represent the average frequency of the system.

The following section will explain the disturbance-based measurement method used for inertia estimation.

6.5   Power system inertia estimation

The kinetic energy (E_k) stored in the rotating mass of a synchronous generator (SG) is defined at the nominal frequency (f_n) as:

$$E_k = \frac{1}{2}J\omega_n^2 \tag{6.1}$$

where J is the moment of inertia and ω_n is the angular speed in rad/s.

The response of a synchronous machine can be described by a parameter called the inertia time constant, which is the ratio of stored kinetic energy to the machine's rated apparent power. The stored kinetic energy can be calculated using (6.1) and the rated apparent power is given in MVA. The mathematical expression for the inertia time constant (H) is used to represent this parameter is given by:

$$H = \frac{J\omega_n^2}{2S_n} \tag{6.2}$$

The inertia time constant (H) characterizes the response of a synchronous generator (SG) and represents the maximum duration in seconds that the machine can supply its rated active power to the power system without any additional mechanical power input until it comes to a complete stop. The value of H depends on the type and size of the SG and ranges from 2 s to 10 s [11]. A larger rated power generally results in a larger inertia constant for a given SG type [2]. Typical or average values for the inertia time constant (H) of coal-fired, gas-fired, nuclear, and hydraulic synchronous generators (SGs) are 3.5 s, 5 s, 4 s, and 3 s, respectively. The values of H for each type of SG can vary depending on factors such as their size and design, but these typical or mean values provide a useful reference point for evaluating the response characteristics of different types of SGs. Understanding the range of H values for different types of SGs is important for developing effective control strategies and ensuring the stable and reliable operation of power systems. [12].

In case of small frequency variations, the dynamics of the rotor of an SG can be represented using the following form of the swing equation.

$$2H\frac{d\Delta\omega(t)}{dt} = \Delta p_m(t) - \Delta p_e(t) - D\Delta\omega(t) \tag{6.3}$$

$\Delta\omega(t)$ is the difference between the actual speed of the rotor and the synchronous speed in p.u. and Δp_m and Δp_e are the sudden change in the mechanical and electric powers, respectively. D is the damping factor of the SG.

Immediately after a disturbance, it is assumed that the mechanical power remains constant, which implies that $\Delta pm = 0$. The deviation of rotor speed ($\Delta\omega$) can be expressed as the rotor frequency deviation which can be approximated as $\Delta f(t)$ [11]. In this context, $\Delta f(t)$ indicates the frequency deviation at the SG's connection bus. By using these assumptions, the swing equation presented in (6.3) can be reformulated as below:

$$2H\frac{d\Delta f(t)}{dt} = -\Delta p_e(t) - D\Delta f(t) \tag{6.4}$$

By using the center of inertia (CoI) concept [14], the swing equation (6.4) can be expanded to describe the frequency dynamics of multimachine power systems. In this approach, all synchronous generators in the studied area are combined into a single equivalent unit, and the swing equation (6.4) is modified accordingly [15]:

$$2H_{tot}\frac{d\Delta f_{sys}(t)}{dt} = \Delta p_{m,tot}(t) - \Delta p_{e,tot}(t) - D_{tot}\Delta f_{sys}(t) \tag{6.5}$$

where $\Delta p_{m,tot}$ and $\Delta p_{e,tot}$ are the total change in the mechanical power and the total change in the electric power change of all the SGs in the system, respectively. D_{tot} is the total damping factor. H_{tot} is the total (overall) inertia constant of the complete power system. H_{tot} can be defined as in (6.6), while f_{sys} (the CoI frequency) can be calculated based on (6.7) [16]:

$$H_{tot} = \frac{\sum_{k=1}^{N} H_k S_k}{\sum_{k=1}^{N} S_k} \tag{6.6}$$

$$f_{sys} = \frac{\sum_{k=1}^{N} H_k S_k f_k}{\sum_{k=1}^{N} H_k S_k} \tag{6.7}$$

where N is the total number of the SGs connected to the system. H_k, S_k, and f_k are the inertia constant, total apparent power, and frequency of the kth SG, respectively.

Due to the need for inertia in modern power systems and advancements in the control of IBRs, the concept of virtual inertia has emerged. By controlling the inverter to behave like synchronous machines, virtual inertia can be added to the power system. This can be beneficial because virtual inertia can be used to mitigate issues arising from high penetration levels of renewable energy sources (RESs). If the inverters have this feature, (6.7) can be modified to combine the inertia effect due to the synchronous generators and the virtual inertia effect due to IBRs as [12]:

$$H_{tot} = \frac{\sum_{k=1}^{N} H_k S_k + \sum_{k=1}^{N} VH_j S_{Vj}}{\sum_{k=1}^{N} S_k} \tag{6.8}$$

where VH_j and S_{Vj} are the virtual inertia constant in (s) and rated power of the jth virtual synchronous machine, respectively.

In situations where the value of the inertia constant is unavailable, or where the power system contains several IBRs, it may be necessary to estimate the system's inertia following a disturbance event. In such cases, a commonly employed equation for the estimation of inertia is given below:

$$H_{est} = \frac{-\Delta P}{2 \times \frac{df}{dt}} \tag{6.9}$$

where ΔP is the active power imbalance.

6.6 Case study of a power system with integrated wind energy plant

6.6.1 Frequency response to different IBR integration levels

To investigate the impact of IBRs on system inertia, a three-bus power system with three synchronous machine (SM)-based power plants is simulated in this study (as shown in Figure 6.2). Each synchronous machine in the system is assumed to have 8.8 s of inertia. The study aims to examine the impact of integrating wind farms into the system by replacing one of the SMs connected to Bus 2, in stages.

Initially, a 20 MVA wind farm is integrated, replacing 20 MVA of the SM connected to Bus 2. Subsequently, the wind farm capacity is increased in steps of 20 MVA (i.e., 40 MVA, 60 MVA, and 80 MVA), which corresponds to wind energy integration levels of 0%, 10%, 20%, 30%, and 40%, respectively.

To analyze the impact of wind energy integration on system inertia, the system frequency response to a sudden increase in load at Bus 1 is monitored. The load was increased by 3.5 MW at time $t = 400$ s. The results of the study are shown in Figure 6.3, which clearly demonstrates the impact of wind energy integration on system inertia.

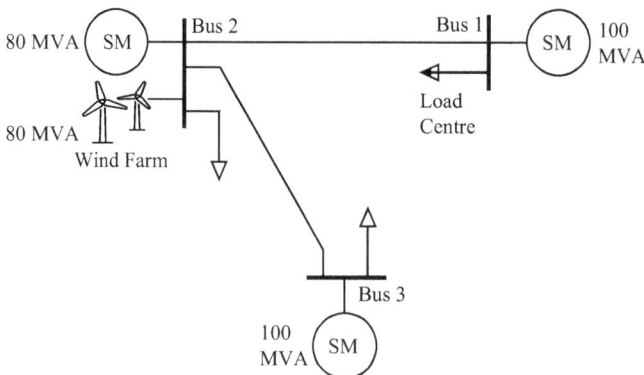

Figure 6.2   A three-bus power system

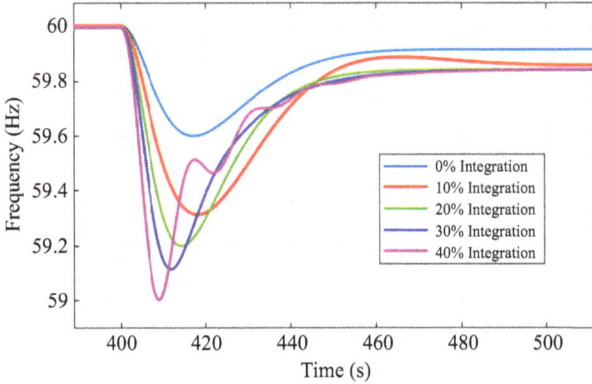

Figure 6.3   *Frequency response to a sudden load increase at different wind energy integration in the power system is shown in Figure 6.2*

The results indicate that as the wind energy integration level increases, the RoCoF and Nadir values also increase. For instance, when there is no wind energy integration (0% integration), the RoCoF is minimal, which is indicated by the slow response of the curve, and the Nadir value is 0.4 Hz. However, when the wind energy integration level is 40%, the RoCoF value is maximum, which results in a rapid change in frequency, and the Nadir value of about 1 Hz. Such a high Nadir value may trigger the under-frequency protection in the system, which can potentially lead to a blackout. Therefore, it is crucial to consider IBRs to maintain system inertia and avoid frequency stability issues when integrating renewable energy sources like wind farms into the grid.

6.6.2   System inertia estimation at different IBR integration levels

This section demonstrates the estimation of the inertia based on (6.9). Initially the inertia is estimated with only the SM machines connected to the power system in Figure 6.2 and compared the results with the calculation based on (6.7) for validation. Then the estimation method is used to estimate the inertia with 40% integration.

Figure 6.4 shows the frequency response and RoCoF at 0% integration. The RoCoF curve shows a maximus value of RoCoF of 0.0407 Hz/s. The frequency response also shows a Nadir value of 0.4 Hz in this case. Thus, the estimation of the inertia can be illustrated as below:

$$\Delta P = 3.5 \text{ MW} = \frac{3.5 \text{ MW}}{100 \text{ MVA} + 100 \text{ MVA} + 80 \text{ MVA}} = 0.0125$$

$$\frac{df}{dt} = \frac{0.0407}{60} = 6.7833 \times 10^{-4}$$

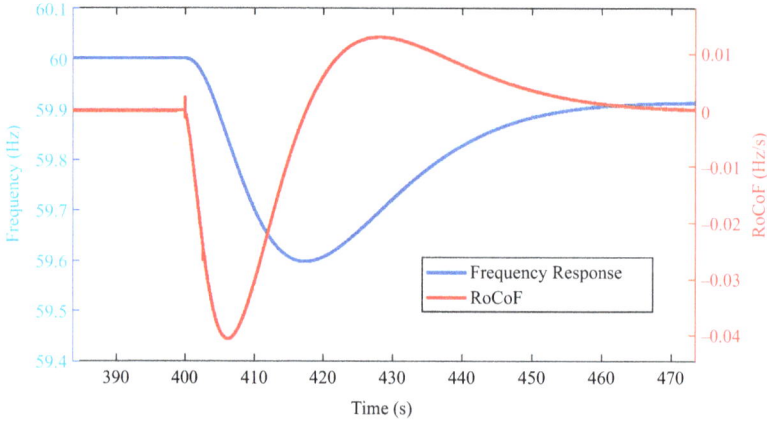

Figure 6.4   Frequency response due to load step with the RoCoF in case no wind integration

Thus, substituting in (6.9) to estimate the inertia as

$$H = \frac{0.0125}{2 \times 6.7833 \times 10^{-4}} = 9.2s$$

To validate this value, (6.7) is used based on knowing the inertias of all of the SM in the system. In this case, all the SMs have been simulated with $H = 8.8$ s. Thus, using (6.7), we get

$$H_{tot} = \frac{8.8 * 100 \text{ MVA} + 8.8 * 100 \text{ MVA} + 8.8 * 80 \text{ MVA}}{100 \text{ MVA} + 100 \text{ MVA} + 80 \text{ MVA}} = 8.8 \text{ s}$$

Though there is an error of about 4.5% between the estimation and the calculation, the estimation method still can be used in case of the integration of the IBRs. Below demonstrates the estimation in the case of 40% integration of wind in the system in Figure 6.2. The response of the frequency and the RoCoF in this case are shown in Figure 6.5.

For the same $\Delta P = 0.0125$ in the previous case, the Nadir value in this case is 1 Hz as can be seen from the figure.

From Figure 6.5, the maximum value of RoCoF is shown as 0.181 Hz/s. The value of df/dt in this case is calculated as

$$\frac{df}{dt} = \frac{0.181}{60} = 3.017 \times 10^{-3}$$

Thus, substituting in (6.9) to estimate the inertia as

$$H = \frac{0.0125}{2 \times 3.017 \times 10^{-3}} = 2.1 \text{ s}$$

This value of H indicated a significant drop in the inertia with 40% integration.

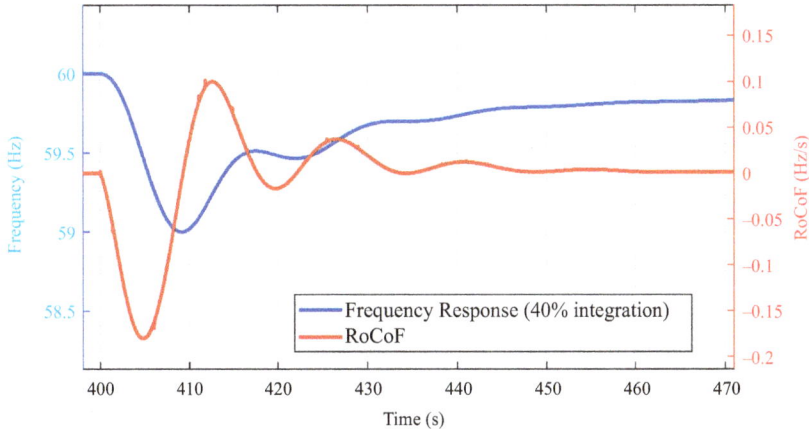

Figure 6.5   Frequency response due to load step with the RoCoF in case no 40% wind integration

6.7   Research gaps, industry challenges, and future research directions

6.7.1   Research gaps

Though there has been significant research on the impact of IBRs on power system inertia, there are still several research gaps that need further investigations. Some of these gaps can be summarized below:

- There is a need for an enhanced understanding of the exact impact of IBRs on system inertia. While it is recognized that IBRs generally have low or zero inertia, there is still a lack of comprehensive understanding of the exact magnitude of the impact of IBRs on system inertia. More research is needed on inertia estimation to quantify the extent to which IBRs contribute to the reduction in system inertia and how this reduction varies with different types of IBRs, their penetration levels, and system configurations.
- There is a lack of standardized methodologies for assessing IBR impact on system inertia. There is a need for standardized methodologies to accurately assess the impact of IBRs on system inertia. Currently, various approaches are used in different studies, which can lead to inconsistent results and make it challenging to compare findings across different research efforts. Developing standardized methodologies for assessing IBR impact on system inertia would help in creating a consistent framework for research and industry practices.

6.7.2   Industry challenges

The issue of inertia in power systems is a relatively new challenge for the power system industry, and while some countries, such as Australia, have taken measures to address this issue, there are still several challenges to be considered for future power systems. Some of these challenges can be listed below:

- Integrating IBRs into existing power systems can be challenging due to issues such as grid code compliance, technical requirements, and coordination with existing grid infrastructure. Utilities, system operators, and grid operators face challenges in managing the impacts of IBRs on system inertia, including potential frequency and voltage instability, as well as the need for updated control strategies and protection schemes to accommodate the changing dynamics introduced by IBRs.

- As the penetration of IBRs increases in power systems, there is a need for effective planning and operation strategies to ensure system stability and reliability. The dynamic behavior of power systems with high IBR penetration can be complex, and there may be challenges in predicting and managing the impacts of IBRs on system inertia, frequency response, and voltage stability. Utilities and grid operators need to develop new approaches and tools to address these challenges.

6.7.3   Future research directions

In the short term, research directions may focus on addressing the research gaps discussed in Section 6.7.1. However, in the long term, as power systems continue to accommodate IBRs, power system operation and control will need to adapt to the new characteristics and dynamics introduced by IBRs. As power systems will be decoupled from the main source of energy by inverters in the IBRs, the coupled relationship between frequency response and inertia may disappear. As a result, future power system may not have a single frequency, but rather can be areas of several frequencies. Therefore, future research directions could include investigating new approaches to system operation and control that account for the changing dynamics of IBRs, as well as developing advanced control strategies, protection schemes, and planning tools that effectively integrate IBRs into power systems while maintaining stability, reliability, and resilience.

6.8   Conclusions

In summary, the large-scale integration of IBRs has impacted the operation of power systems in various ways, with two major impacts being the reduction in system inertia and power system strength. Traditionally, the rotating masses of synchronous generators have been providing sufficient inertia to maintain frequency stability in power systems. However, as more IBRs replace synchronous generators, the issue of low inertia has emerged, making power systems more vulnerable to instability. Similarly, the retiring of synchronous generators and their replacement with IBRs has reduced the fault current contribution, which has been a key measure of system strength. As a result, the strength of power systems is declining, making them more vulnerable to voltage instability.

To address these challenges, several countries have already introduced updated requirements for power system inertia and system strength, which reflect the changing dynamics of power systems. Furthermore, research has been conducted and several methods have been introduced to measure system inertia which have been discussed in the previous sections. However, further research is needed to better understand the extent to which IBRs contribute to the reduction in system inertia and how this reduction varies with different types of IBRs, their penetration levels, and system configurations.

References

[1] "Tracing the London Blackout," in *Power Engineer*, vol. 17, no. 5, pp. 8–9, 2003, https://digital-library.theiet.org/content/journals/10.1049/pe_20030501; jsessionid=30cb0p0hk29ci.x-iet-live-01.

[2] O.P. Veloza and F. Santamaria. "Analysis of major blackouts from 2003 to 2015: classification of incidents and review of main causes." *Electricity J.* 2016, 29(7), 42–49.

[3] R. Yan, N.-A. Masood, T. Kumar Saha, F. Bai, and H. Gu. "The anatomy of the 2016 South Australia blackout: a catastrophic event in a high renewable network." *IEEE Transact. Power Syst.* 2018, 33(5), 5374–5388.

[4] Hurricanes Nate Maria Irma and Harvey Situation Reports, Department of Energy, Aug. 2018, https://www.energy.gov/ceser/downloads/hurricanes-nate-maria-irma-and-harvey-situation-reports.

[5] N.M. Flores, H. McBrien, V. Do, M. V. Kiang, J. Schlegelmilch, and J. A. Casey. "The 2021 Texas power crisis: distribution, duration, and disparities." *J. Exposure Sci. Environ. Epidemiol.* 2023, 33.1, 21–31.

[6] ENTSO-E, "High Penetration of Power Electronic Interfaced Power Sources and the Potential Contribution of Grid Forming Converters." untitled-292051-ea.pdf (euagenda.eu).

[7] B. Porretta and S. Porretta. "Calculation of power systems inertia and frequency response." In *2018 IEEE Texas Power and Energy Conference (TPEC)*, College Station, TX, 2018, pp. 1–6.

[8] J. Bryant, R. Ghanbari, M. Jalili, P. Sokolowski, and L. Meegahapola. "Frequency Control Challenges in Power Systems with High Renewable Power Generation: An Australian Perspective." RMIT University, 2019.

[9] National Grid ESO. Electricity System Operator Grid Code. Section 5: Operational Security Standards, 2018. https://www.nationalgrideso.com/document/122101/download.

[10] Australian Energy Market Operator. *Inertia Requirements Methodology: Inertia Requirements & Shortfalls.* Australian Energy Market Operator (AEMO): Melbourne, Australia, 2018, p. 46.

[11] S. C. Dimoulias, E. O. Kontis, and G. K. Papagiannis. "Inertia estimation of synchronous devices: review of available techniques and comparative

assessment of conventional measurement-based approaches." *Energies* 2022, 15.20, 7767.

[12]   K. Prabhakar, S. K. Jain, and P. K. Padhy. "Inertia estimation in modern power system: a comprehensive review." *Electric Power Syst. Res.* 2022, 211, 108222.

[13]   D. Zografos, M. Ghandhari, and R. Eriksson. "Power system inertia estimation: utilization of frequency and voltage response after a disturbance." *Electric Power Syst. Res.* 2018, 161, 52–60.

[14]   A. Gorbunov, J. C. H. Peng, J. W. Bialek, and P. Vorobev, "Can center-of-inertia model be identified from ambient frequency measurements?" *IEEE Trans. Power Syst.* 2022, 37, 2459–2462.

[15]   B. Tan, J. Zhao, M. Netto, V. Krishnan, V. Terzija, and Y. Zhang. "Power system inertia estimation: review of methods and the impacts of converter-interfaced generations." *Int. J. Electr. Power Energy Syst.* 2022, 134, 107362.

[16]   S. Azizi, M. Sun, G. Liu, and V. Terzija. "Local frequency-based estimation of the rate of change of frequency of the center of inertia." *IEEE Trans. Power Syst.* 2020, 35, 4948–4951.

Chapter 7

The effect of power system strength on the calculation of available transmission capacity

Mostafa Eidiani[1]

Abstract

This chapter provides an overview of past ideas and perspectives and future visions of available transmission capacity (ATC). Methods can be classified into two general categories: dynamic and static, and the methods presented compete in both categories according to their accuracy and speed. This chapter introduces ATC, reviews conventional definitions, concepts, and terms, and, finally, evaluates the disadvantages and advantages of each method. Also, it is discussed why holomorphic algorithms are advantageous when determining ATC. ATC can also be determined faster and more accurately using "differential equation load flow." Due to the presence of HVDC lines and wind power plants, ATC calculations are challenging. The following sections explore the problems associated with inaccurate power network information and the use of state estimation in ATC calculations. Lastly, cyber-attacks to change ATC data are examined and their impact on the electricity market is discussed. The methods provided in this chapter will be very useful to future ATC researchers for developing faster and more accurate methods.

Abbreviations

AI	artificial intelligence
ANRS	approximate Newton–Raphson–Seidel
ANRS-DDH	approximate Newton–Raphson–Sidel and developed Down-Hill algorithm
APEBS	approximation of the potential energy boundary surface
ATC	available transmission capacity
BDD	bad data detection
BE	bee algorithm
CBM	capacity benefit margin

[1]Energy Security and Sustainable Energy Institute, Iran

C-DATC	combined approach for dynamic ATC
CPF	continuation power flow
CPPS	cyber-physical power system
CSA	cuckoo search algorithm
CTSA	complex transient stability analysis
CUEP	controlling unstable equilibrium point
CUSUM	cumulative sum algorithm
DATC	dynamic available transmission capacity
DEA	differential evolution algorithm
DE	differential equation
DELF	differential equation load flow
DPF	dynamized power flow
DG	distributed generation
DH	Down-Hill algorithm
DDH	developed Down-Hill algorithm
ETC	existing transmission commitment
FADATC	fast and accurate ATC
FCTTC	first contingency total transfer capability
GWO	Grey Wolf Optimizer
GMRES	general minimum residual method
HELF	holomorphic embedding load flow method
ICT	information and communication technology
IoT	Internet of Things
LF	load flow
MD	minimum distance
NR	Newton–Raphson
NRS	Newton–Raphson–Seidel
OPF	optimal power flow
PATC	probabilistic ATC
PEBS	potential energy boundary surface
POMP	point of maximum potential
PMU	phasor measurement unit
RHELF	revised method of HELF
RMSE	root mean square error
RSC	relative speed of calculations
SATC	static available transmission capacity
SCADA	supervisory control and data acquisition
SC-OPF	security-constrained optimal power flow

SCR	short-circuit ratio
SE	state estimation
SVR	support vector regression
TRM	transmission reliability margin
TSC	transient stability constrained
TTC	total transfer capability
UEP	unstable equilibrium point
WLS	weighted least squares

7.1 Introduction

Today's power systems are affected by almost every aspect of sustainability, including competition at different levels [1,2]. Increasing the density of transmission lines will maximize profits, but there will also be problems with voltage stability and other constraints that threaten the system's safety [3]. It is vital to estimate and calculate available transmission capacity (ATC) [4].

Two general categories of ATC determination are available: static and dynamic. In static models, system variables do not change over time, but in dynamic models, dynamic and static variables are considered together. In determining ATC, stopping criteria prevent transmission power increases. Transient and voltage instability, bus voltage limits, static stability limits, AC/DC load flow divergence, and thermal limits of lines are some of these limitations [5–7].

This introduction is divided into several sections based on the amount of information it contains. Section 7.1.1 briefly explains power system strength basics. Additionally, there is a short section discussing challenges and future research directions. Various concepts and definitions of ATC are discussed in Section 7.1.2. Abbreviations and their relationship to one another are very helpful for readers to know. Section 7.1.3 presents relevant static ATC (SATC) methods by date of presentation and importance. In the following section, we will discuss the most significant methods. Sections 7.1–7.4 present important dynamic ATC (DATC) methods by date of presentation and importance. In the following section, we will discuss the most significant methods.

There is a brief explanation of the methods (Holomorphic approach), (Holomorphic hybrid method), (Approximate NRS algorithm), (Developed DH algorithm), (Revised method of Holomorphic embedded load flow), and (approximate potential energy boundary surface method (APEBS) in the second part.

The third part discusses how differential equation load flow (DELF) plays a role in determining ATC in static and dynamic scenarios. The fourth section discusses the effect of HVDC lines and wind power plants on ATC. ATC determination is illustrated in the fifth section, which involves state estimation, information failure, and the uncertainty effect. Cyber security in the power system and the effect of weighted least squares (WLS) on determining ATC are presented in the sixth section. The following are general concepts and definitions (ATC).

7.1.1   *The basics of power systems strength*

There have been no problems with traditional power systems' stability and security for decades. In response to environmental challenges and energy security risks, modern power systems based on smart grid concepts are inevitable [8]. System strength and inertia are essential for secure power systems, which are provided by synchronous power generation. Power system strength is its ability to maintain constant voltage levels despite fluctuations in demand and supply. Modern power systems depend on system strength and inertia to maintain acceptable stability levels. Power system instability and supply interruptions are more likely to occur if there are not enough of these services.

In power systems strength measurement, the short-circuit ratio (SCR) and the ratio of reactance to resistance (X/R) are usually used [9,10]. In modern power systems, the SCR is no longer a valid indicator of system strength [11]. In [12], a detailed definition of the SCR is examined, as well as factors affecting its applicability to converter-dominated power systems. Operators around the world are searching for new metrics to evaluate power systems. Modern systems can increase green and renewable energy hosting capacity, accelerating the decarbonization plan.

In general, "System Strength" describes a broad and complex area of generator and system stability, protection and normal operability, system economics, and power quality [10]. In theory, system strength refers to the electric power system's ability to maintain voltage during and after reactive current injection. After reactive power injection, a grid with high system strength can experience fewer changes than a grid with low system strength.

The Australian energy market operator defines system strength as the ability of a power system to maintain stable operating conditions in both steady-state and after a disturbance [10]. With the development of system strength, it is now used to describe how sensitive any power system is to maintaining and controlling voltage.

In the electricity market, system strength has historically been produced as a byproduct of large synchronous generators. In transitioning to a low-carbon future, ensuring system strength is becoming increasingly critical due to the rapid connection of non-synchronous generation sources. Power system controllability, reliability, and stability have been challenged by the high penetration of asynchronous renewable energy.

Power system strength is challenged due to the increasing integration of renewable energy resources (RERs) into the electric grid through power electronic interfaces. Inverter-based RERs (IB-RERs) may contribute to weak grid conditions due to their high penetration. Renewable energy generation uncertainty makes it challenging to assess grid strength. The probabilistic collocation method quantifies grid strength probabilistic characteristics under uncertain renewable generation [13].

The authors of [14] examine PV's impact on fault levels as an indicator of network strength. The sync condenser's voltage and reactive power can be independently controlled to support the system fault levels. System strength has been increased by using synchronous condensers rather than more synchronous generators.

According to the above description, there is a need for comprehensive research on power system strength that includes applications, case studies, best practices, and the latest evaluation methods.

7.1.2   Concepts and definitions of ATC

ATC is briefly discussed in this section so that ATC methods can be better understood [15]:

- Transfer capability: Power can be transferred from one region to another while maintaining network limitations.
- Transfer capacity: Network equipment bearing capacity.
- Criteria for stopping: Transient and voltage instability, bus voltage limits, static stability limits, AC/DC load flow divergence, and thermal limits of lines.
- Destination in determining ATC: group to group, bus to group, group to bus and bus to bus.
- Parallel paths: Power is also transmitted indirectly between two regions.
- Simultaneous and non-simultaneous ATC: When there is no power exchange between two areas, power can be transferred between them.
- Total transfer capability (TTC): Contracting limits, stability, voltage, and thermal are all measured by TTC.
- Transmission reliability margin: To be secure, a system must always retain a certain amount of ATC.
- Capacity benefit margin: To ensure load-supplying organizations receive reserve power, CBM is measured as part of ATC.
- Operating horizon: Describes planning from now until 31 days in the future.
- Planning horizon: A plan for 10 years or 1 year in the future.
- Non-interruptible and interruptible planning and reserve: Transfer services include non-recallable and recallable scheduled and reserved.
- Existing commitment: A contract can be terminated under an EC whether it is long-term or short-term.
- Available transfer capability: TTC minus CBM minus EC minus TRM.
- First contingency total transfer capability (FCTTC): After one component of the FCTTC system is disconnected, the TTC remains stable [16].
- Recallable available transfer capability (RATC): NATC minus recallable transmission service (RTS).
- Non-recallable available transfer capability (NATC): TTC minus non-RTC (including CBM) minus TRM.
- Principle 1: Electricity market power transferability should be determined by ATC calculations.
- Principle 2: In calculating ATC, non-simultaneous and simultaneous power transmission must also be considered.
- Principle 3: ATC should be calculated with all the necessary information.
- Principle 4: Information specifying the ATC value of the transmission network must be sent and received through regional coordination.

- Principle 5: ATC must meet several criteria, including system reliability plans, operating policies, and public and private exchange systems.
- Principle 6: In determining ATC, it is essential to consider reasonable uncertainties in system variables.
- Principle 7: To calculate ATC, transmission providers must take into account all system conditions.
- Principle 8: ATC coordination and sending network information are region members' responsibilities.
- Principle 9: ATC should also be available in other areas and to customers.
- Principle 10: Customers should receive updates on transmission routes' ATC from transmission providers.
- Principle 11: A limited number of routes are used to send information monthly.
- Principle 12: ATC is sent daily, the next day, and 6 days later on unlimited routes.

7.1.3 Static ATC

There are many methods of calculating ATC (SATC), but the static method is the oldest. The most commonly used methods include sensitivity or linearization calculations, an optimal power flow [17] calculation, and a load flow calculation. The first classical continuous power flow (CPF) method was developed by Ajarapo [18] to investigate voltage stability in a static state. In 1998, Sauer outlined possible methods for reducing ATC impact on transmission loading relief (TLR) and improving transmission system efficiency [19]. The author of [20] says ATC can be calculated based on thermal limits, voltage collapse, voltage limits, reactive power effects, and AC load flow.

It is possible to determine the ATC between two or more regions using Hamud's method [21]. CPF is suggested to be improved using the general minimal residue (GMRES) method in their study [22]. The problem has become more complicated with the improvement of reactive power and wind power plants as well as the determination of voltage stability [23]. As an alternative to Newton–Raphson (NR), Newton–Raphson–Seydel (NRS) has been developed [24], which is faster and more accurate. ATC and voltage stability can be determined more accurately and quickly using this method than previous methods [25]. Due to the introduction of a new method called minimum distance (MD) [26,27] in 2011, it has been used to determine ATC and voltage stability, in which the distance between the operating point and the system's stability limit is estimated with sufficient accuracy and speed as consumption and production increase. In addition, the author of [28] presents an approach to simultaneously detecting transmission and distribution network voltage stability. The Jacobian matrix calculation is faster and more accurate by combining CPF-GMRES, NRS, and Down-Hill and removing trigonometric parameters [29]. In [30], authors used Cat optimization algorithms and SATC to locate FACTS tools optimally. Data-driven sparse polynomial chaos expansion was introduced in the paper [31] to evaluate probabilistic TTCs including renewable energy sources.

Recent academic research [32,33] has shown that the dynamized power flow (DPF) approach works well for simulating transient stability. Ref. [34] shows that nonlinear AC load flow became linear after differential transformation.

The differential load flow equation (DELF) has been improved using the inverse matrix approach [35] to make power flow models more accurate when applied to SATC evaluations. Milliapalas showed that an appropriate probabilistic method could be used to calculate ATC [36]. Due to nonlinear programming, ATC can be calculated and influenced to maximize transmission power [37].

Sauer's paper [38] calculates ATC using reactive power. IEEE systems calculate TTC using a curve analysis algorithm from [39]. Based on small-signal stability constraints, Ref. [40] shows that nonlinear programming methods cannot be utilized to determine SATC.

The calculation of optimal power flows has changed in recent years due to artificial intelligence (AI) techniques [41]. ATC is determined using various AI methods, including GA [42], CSA [32], BA, PSO, ANN, and GWO. Based on 2020 [33], CSA is more appropriate for determining ATC for the electricity market than GWO and PSO iterative methods. However, AI methods are fast but consume a lot of information, increasing the problem of choosing the right training method.

Ultimately, static methods are simple, inexpensive, clear, flexible, and fast since differential equations are not needed, and there is no need to use differential equations. Accuracy and optimism are also disadvantages.

7.1.4   Dynamic ATC

An ATC with dynamic constraints (DATC) combines constraints from differential and algebraic equations (DAE). DAE problems are very difficult to solve. Static equations can be converted into dynamic equations to resolve the problem [34,43]. DAE equations cannot be solved faster than ordinary differential equations.

In 1992, the authors presented a method for classifying and screening a large number of events based on transient stability limits. However, the method failed to take nonlinear loads into account [44]. Paula's study shows that this method is advantageous because it considers both the dynamic and static limits of transient stability. However, it fails to take bus voltage into account when computing ATC using event ranking in the base state [45]. Authors calculated DATC using two novel methods in 2002 and demonstrated that the presented method is accurate and fast [46]. Xingbo calculated dynamic ATC in 2004 based on uncertainty, transient and voltage stability, and thermal limits [47]. In 2007, Junji developed a dynamic ATC for a small system [48], resulting in optimal load flow.

Venkaiah invented the DATC method in 2008 [49], which is faster and more accurate than neural networks. The researchers tested the PEBS method against error backpropagation learning (BPA), radial basis function learning (RBF), and adaptive neural fuzzy inference (ANFIS), and found that ANFIS had the fastest response time and the fewest errors, but required a long training process [50].

The support vector regression (SVR) method was presented by Shirinvasan in 2013 [51] and 2015 [52]. By using the PEBS method, he determined different load patterns in the SVR. It was shown that different evolutionary algorithms and SVR performed much better than neural networks in 2017 [53].

In 2018, Shaban calculated the controllable unstable equilibrium point and transient stability limit using a fast ATC considering voltage and output constraints through a quadratic approximation [54]. It discovered the exact solution without initial estimation using holomorphic embedding powers without initial estimation [55,56]. ATC can be calculated faster and more accurately by combining several old methods, such as [57].

Using differential power/load equation flow (DEPF-DELF) to create fictitious dynamic models from power flow models is a novel approach. A DEPF analyzes the whole time domain in a dynamic manner instead of repeatedly resolving power flow under specific conditions. DATC calculations were conducted using a hybrid method based on DELF and minimum distance [58].

Similar to SATC determination methods, it was necessary to use state estimation instead of normal load flow to compensate for inaccurate or incorrect information. In [59,60], these cases have been evaluated using state estimation and old methods of ATC determination and DIgSILENT PowerFactory.

In Ref. [4], cyberattacks are simulated in DIgSILENT for state estimation and energy management. ATC speed and state estimation are improved by weighted least squares in the newly developed method.

In [61], HVDC lines, large loads, and wind farms were examined in ATC calculations. Attempts were also made to improve the previous technique by using boundary surface potential energy estimation.

Based on DATC determination papers, smart methods are becoming more popular, but the issue of training time and how to use smart methods remains. Dynamic methods have the following advantages over static methods: There are more details, the possibility of more corrective actions, the possibility of applying the voltage–frequency relationship, the possibility of considering larger changes, and finally, more accurate results.

The following should be considered for future DATC methods:

- Online and real-time completion of DATC should be possible.
- Future power systems should be designed using a probabilistic approach, since electric vehicles [62–64], solar power plants [65], hydrothermal and pumped-storage units [66], and distributed resources will not be possible without them.
- Data management is essential for intelligent methods to work.
- Power systems will be decentralized in the future. The current method of managing and controlling energy is not suitable for such systems, so ATC determination methods have to change as well.

ATC is explained in more detail in the following section based on the most relevant methods in recent years.

7.2   DATC and holomorphic approach

This section describes how holomorphic embedded load flow (HELF) relates to ATC [55–57]. Non-iterative HELF [55–57] uses concepts within complex analyses without requiring any simplified assumptions or initial values. It is evident from this comparison that HELF has many shortcomings, particularly in terms of computation time.

The present section uses the approximate Newton–Raphson–Seidel (ANRS) algorithm and the Down-Hill algorithm (DDH). ANRS-DDH and the revised holomorphic embedded load flow (RHELF) were combined to take advantage of both methods. APEBS-based methods for determining transient stability were also used to correct previous methods and approximate the potential energy boundary surface (APEBS). DATC (C-DATC) computation is based on a combination of static and transient stability approaches. An analysis of these four methods shows that a small correction improves the speed and accuracy of the proposed method [55–57].

The proposed methods for ATC calculation should not only exhibit better accuracy and speed but also exhibit the following characteristics.

• Online computation of DATC is recommended.
• Contracts between districts should also be included in DATC.
• DATC should be computed and expanded using probabilistic methods as distributed generation (DG) will be a feature of all power systems in the future.
• Artificial intelligence methods should be made more efficient by data management.
• Since centralized ATC methods are obsolete in the future, decentralized systems will replace them soon.

References [55–57] have the following salient features:

1. Wind farms with large-scale transfer capabilities are analyzed using the proposed method.
2. Under any conditions, the proposed method is stable, robust, accurate, and fast.
3. Four improved methods are combined in the proposed method.
4. Using the proposed method, acceptable results have been obtained for a large real network with a large number of wind farms.
5. An innovative approach was developed to improve transient stability and load flow in dynamic ATC by combining APEBS, RHELF, DDH, and ANRS methods.
6. To make the proposed method more accurate and faster, a small correction has been applied to these four methods.
7. A summary of the DATC hybrid designation method and the new load flow program is presented in the following section.

7.2.1 DATC and holomorphic hybrid method

In dynamic and static power systems, load flow is a key component. Using the approximate NRS (ANRS) procedure, it is possible to determine stable and unstable points and the limits of stability [22,28]. NRS may not be the most suitable approach for distribution networks. As a means of extending the convergence region of the approach, the Down-Hill (DDH) method (Sections 7.2–7.6) is used in this section.

The Jacobian matrix's inverse calculation time increases exponentially as its dimensions increase. Inverting the submatrices rather than inverting the Jacobian matrix directly can significantly reduce the inverse calculation time (Sections 7.2–7.6). Reducing the inverse calculation time of the Jacobian matrix directly impacts DATC and ATC and load flow calculations. Based on the initial values, iterative methods are used to solve this non-linear load flow problem. It uses complex analysis concepts, rather than initial values or simplification assumptions, to achieve its non-iterative approach (Sections 7.2–7.7).

In addition to finding the most accurate operable solution, RHELF provides an unambiguous signal. Therefore, RHELF assists in calculating accurate results without taking assumptions into account or starting from a predetermined position.

The authors of [55–57] review HELF's shortcomings regarding computation time. An approach is presented here that combines Down-Hill (DDH) and Newton–Raphson–Seidel (ANRS) with the RHELF algorithm to produce an algorithm that is more accurate and faster on average than iterative methods and does not depend on initial values.

ANRS-DDH-RHELF uses a novel approach to combine Down-Hill (DDH) and approximate Newton–Raphson–Seidel (ANRS) with the advantages of RHELF. It is more accurate and faster than iterative methods on average, and it does not depend on initial values.

Combining RHELF, DDH, and ANRS is as follows:

- The first step is to determine a quick answer to power flow using RHELF without the need for initial values or simplification assumptions.
- A reliable method, accurate, and fast for determining power flow is identified through the combination of DDH and ANRS.
- ANRS-DDH can be calculated using the RHELF method, which provides an accurate and fast answer used for static ATC.

Listed below are the highlights of the proposed SATC evaluation method using (ANRS-DDH-RHELF):

- Due to RHELF's independence from an initial guess, SATC calculations take less time.
- ANRS-DDH's static ATC can be obtained with accuracy and precision.
- ANRS-DDH and RHELF combine to provide faster speed.

The static ATC and dynamic ATC are determined using this updated power flow method in the following. Research in this paper aims at determining DATC in a real network where wind power is generated. Below is a summary of the factors involved:

- Network information is modeled.
- Power transmission contracts determine the place of consumption and production.
- Combining RHELF, DDH, and ANRS, the SATC has calculated an upper bound for DATC (Sections 7.2–7.5). In the first place, RHELF calculates a fast

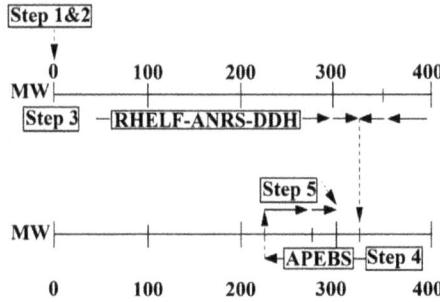

Figure 7.1   Proposed method's main algorithm

power flow answer without any initial values or simplification assumptions. In addition, the RHELF method is used as an initial guess for ANRS-DDH, and the answer is accurate, fast, and can be used for static ATC as well. Except for transient stability, all constraints are considered here.

- By APEBS (approximated PEBS) from Step 3, DATC is determined (Sections 7.2–7.8).
- Using a simulation approach to calculate DATC, a more accurate power system model is developed. When DATC is gradually adjusted, it reaches a stable border.
- Then move back to (2) and apply the network changes and the target information again, and follow the procedure described above again.

Figure 7.1 illustrates the proposed method. The calculation of the energy method based on the voltage stability constraint provides an upper bound on transient stability in ANRS-DDH-RHELF (Sections 7.2–7.8). A four-network algorithm is calculated in accordance with the above conditions.

7.2.2   Example network

A number of these ideas have been implemented in Khorasan networks (850-bus), IEEE 118-bus networks [36], and Iowa State networks [34]. A compact Iranian network of 850 buses is being used (Table 7.1). Due to many wind farms, calculating ATC is further complicated.

7.2.3   Simulation and comparison

This research introduced a method that can be compared to a similar method. During training, intelligent methods require a great deal of information. Determining controlling unstable equilibrium points (CUEPs) is time-consuming [16,54]. Thus, PATC [36], CUEP [54], SVR [53], and FADATC [34] are compared with the speed and accuracy of the novel method.

Table 7.2 compares 11 sales contracts and energy purchases. A 500 MW wind farm and three steam power plants were assumed to generate the power, with sales

Table 7.1   850 bus network information

Title	MVAr	MW	Number	
Generation	−461	2,277	Substations	12
External infeed	16	−683	Lines	395
Load $P(U)$	32	1,527	Syn. machines	54
Load $P(U_n)$	320	1,527	Loads	186
Losses	−19.25	67	Bus bars	459
Line charging	−24.72	–	2-w Trfs	325
Compensation ind.	−24.72	–	Asyn. machines	0
Compensation cap.	11.61	–	Shunts	30
Installed capacity	–	5618	Terminals	391
Spinning reserve	–	3971	3-w Trfs	43

assumed at four grid outlets. Table 7.2 shows that the proposed method has a faster speed than PATC, CUEP, SVR, and FADATC. Under equal conditions, the method showed improved accuracy.

The following equations show the root mean square error (RMSE) and the relative speed of calculations (RSC):

$$RSC_i = \frac{[Method\ i\ Calculation\ time] - [Method1\ Calculation\ time]}{Method\ i\ Calculation\ time} \times 100$$

(7.1)

$$RMSE_i = 1,000\sqrt{\frac{1}{N}\sum_{i=1 \neq 2}^{N}(ATC_{Methodi} - ATC_{Method2})^2}$$

(7.2)

7.2.4   Wind farms In Iran

The ATC has been calculated in the planning mode for the future network. This section considers different scenarios of wind farm output and the additional load arrival. Khorasan networks were used as examples in this section [55–57].

The revised method is used to estimate DATC and SATC values for a number of scenarios, including the entry of loads and wind farms. Several scenarios for the future have been suggested by experts. A summary of the simulation outputs is given in Table 7.3. Table 7.3 shows how the electricity company planned for all possible modes of transportation. Our conclusion in this section is that the proposed approach is ideal not only for operation but also for planning and wind farm existence.

Figure 7.2 shows that the proposed method converges, exact dynamics, approximate dynamics, and SATC.

Table 7.2 Comparison of accuracy and speed of the presented method

Contract	Grid	RSC (Equation (7.2))				Method 6	Method 5	Method 4	Method 3	Method 2	Proposed method
		RSC_3	RSC_4	RSC_5	RSC_6	PATC	CUEP	SVR	FADATC	Simulation	C-DATC
1	Iran	5.5	9.23	22.41	4.41	6.23	6.02	6.21	6.20	6.25	6.24
2		2.80	8.16	12.81	1.12	4.10	3.98	4.09	4.09	4.12	4.11
3		1.44	6.00	11.73	1.38	1.22	1.03	1.20	1.21	1.25	1.24
4		3.76	9.13	No answer	2.21	0.89	–	0.89	0.89	0.92	0.91
5		2.45	7.99	12.63	1.03	4.49	4.21	4.46	4.47	4.48	4.48
6		3.59	8.27	12.27	2.50	2.63	2.45	2.59	2.59	2.61	2.60
7		2.28	7.38	13.02	1.38	0.61	0.47	0.59	0.60	0.62	0.61
8		5.54	10.37	13.89	3.35	3.56	3.35	3.58	3.51	3.54	3.53
9	Iowa State	5.12	11.94	9.57	2.04	1.82	1.68	1.81	1.81	1.83	1.82
10	IEEE 118-bus	5.87	10.64	17.62	4.38	5.19	5.1	5.16	5.18	5.21	5.20
11		8.56	14.82	No answer	4.24	4.87	–	4.83	4.83	4.86	4.85
RMSE		4.26	9.40	12.66	2.55	19.54	168.58	34.38	30.00	0	9.53

Table 7.3　A summary of the simulation outputs

Row	Network		Generation	Load	Loss	DATC	SATC	Simulation
1	Connect all loads	Removal of wind farms	4.852	4.255	0.134	0.159	0.435	0.160
2	Non-connection of 300^{MW} load			3.956	0.131	0.223	0.738	0.222
3	Minimum load of the Khorasan network			3.585	0.152	0.403	1.085	0.404
4	Connect all loads	Connection of 300^{MW} wind farm	4.121	4.255	0.132	0.259	0.734	0.259
5	Non-connection of 300^{MW} load			3.956	0.137	0.409	1.029	0.409
6	Connect all loads	575^{MW} wind farm connection	5.394	4.255	0.140	0.374	0.999	0.373
7	Non-connection of 300^{MW} load			3.956	0.152	0.421	1.287	0.421
8	Non-connection of 427^{MW} load			3.838	0.160	0.649	1.405	0.650
9	Minimum load of the Khorasan network			3.158	0.215	0.765	1.625	0.765

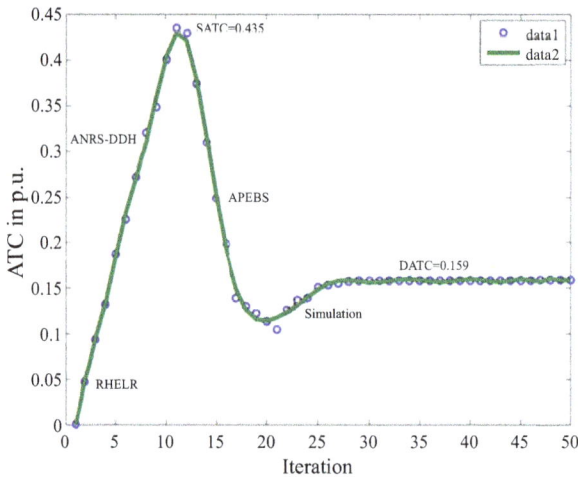

Figure 7.2　Proposed method convergence characteristic

7.2.5   Approximate NRS algorithm

A description of NRS equations can be found at [22,28]:

$$F(X) = \begin{bmatrix} \dfrac{\left(\lambda * I - \dfrac{\partial f}{\partial x}\right) * v}{1 - \|v\|} \\ p * a + -y_0 - f(x) \end{bmatrix} = 0 \Rightarrow \dfrac{\partial F}{\partial X}$$

$$= \begin{bmatrix} \dfrac{\partial^2 f}{\partial x^2} * diag(v_1, v_2, \ldots, v_n) & \dfrac{\partial f}{\partial x} - \lambda * I & 0 \\ 0 & \dfrac{v'}{\|v\|} & 0 \\ \dfrac{\partial f}{\partial x} & 0 & -p \end{bmatrix}, X = \begin{bmatrix} x \\ v \\ a \end{bmatrix} \tag{7.3}$$

Newton–Raphson (NR) and approximate NRS (ANRS) are similar in behavior. With this method, the Jacobian matrices are larger than with the NR method. The saddle-node divergence problem has been omitted. The part below the voltage–power curve (*P–V*) is obtained using this method.

7.2.6   Developed DH algorithm

A new variable ($a_i = \cos\delta_i$) and a revised variable ($b_i = \sin\delta_i$) are defined in the updated load flow equations. Accordingly, in (7.3), the variable *x* changes from (δ,v) to (a,b,v), and the main variable *X* changes from (x,v,a) to (a,b,v,a). In addition to increasing computational speed, removing sine and cosine from the equations makes them a bit more complicated, but it is well worth it. A developed DH algorithm is created using this idea and is called "developed DH" (DDH) [28]. Due to this idea, the proposed approach is more reliable and takes less computation time. As a result, ANRS becomes ANRS+DDH:

$$\Delta X^{(r)} = D_h \times \left[F_X(X^{(r)}) \right]^{-1} F(X^{(r)}) \Rightarrow X^{(r+1)} = X^{(r)} - \Delta X^{(r)} \tag{7.4}$$

The DH algorithm's convergence coefficient is D_h. The Jacobin matrix can be significantly reduced in inverse calculation time by inverting the submatrices rather than directly inverting them. Due to this reduction in the inverse calculation time of the Jacobin matrix, the calculation of load flow, SATC, and DATC are all affected.

Inversion of $F \triangleq \begin{bmatrix} F_{11} & F_{12} \\ F_{21} & F_{22} \end{bmatrix}$ occurs if matrix F and matrix F_{11} are invertible:

$$F_Z \triangleq inv(F_{22} - F_{21}F_{11}^{-1}F_{12}) \Rightarrow inv(F)$$

$$= \begin{bmatrix} F_{11}^{-1} + F_{11}^{-1}F_{12}F_ZF_{21}F_{11}^{-1} & -F_{11}^{-1}F_{12}F_Z \\ -F_ZF_{21}F_{11}^{-1} & F_Z \end{bmatrix} \tag{7.5}$$

7.2.7   *Revised method of holomorphic embedded load flow*

As in [55–57], a summary of the holomorphic embedded load flow approach is presented here. S_i is the complex power at a bus in the n-bus system, where m and p stand for the set of PQ buses and PV buses, respectively. An *i*th bus power flow equation looks like this in the holomorphic form:

$$S_i = P_i + jQ_i = \widehat{V}_i(\sum_{j=1}^{n} Y_{ij}.\widehat{V}_j)^* \Rightarrow \frac{P_i - jQ_i}{\widehat{V}_i^*} = (\sum_{j=1,\neq i}^{n} Y_{ij}.\widehat{V}_j) + Y_{ii}\widehat{V}_i$$

$$\Rightarrow \frac{P_i(s) - jQ_i(s)}{\widehat{V}_i^*(s^*)} = \left(\sum_{j-1,\neq i}^{N} Y_{ij}.\widehat{V}_j(s)\right) + Y_{ii}\widehat{V}_i(s) \tag{7.6}$$

V is the voltage at the bus; Y is the network admittance matrix; s is an embedded complex parameter, and V_1^{sp} is the specified voltage magnitude:

$$P_i(s) = sP_i, \ Q_i(s) = sQ_i, \ V_l(s) = sV_l^{sp} \tag{7.7}$$

$$\left|sV_{sl}^{sp}\right| = |V_{sl}(s)| \tag{7.8}$$

The following is a representation of the power series:

$$X(s) = \sum_{n=0}^{\infty} (s)^n X[n] \tag{7.9}$$

The constant terms and coefficients of (s, s2, s3, ...) on both sides of the equation are compared to determine the value of $V_i[n]$:

$$\sum_{k=1\neq j}^{N} V_k[n] \ Y_{jk} = (W_j[n-1]S_j)^* - V_j[n-1] \ Y_j \tag{7.10}$$

$$\sum_{k=1\neq l}^{N} V_k[n] \ Y_{lk} = Rh_l[n-1] - jQ_l[n] \tag{7.11}$$

which

$$Rh_l[n-1] = W_l^*[n-1] \ P_l - V_l[n-1] \ Y_l - j\sum_{k=1\neq l}^{n-1} W_l^*[n-k] \ Q_l[k] \tag{7.12}$$

$$V_{l\,re}[n] = 0.5\delta_{n1}(V_l^{sp^2} - 1) + \delta_{n0} - 0.5\sum_{k=1\neq l}^{n-1} V_l^*[n-k]V_l[k] \tag{7.13}$$

It is important to understand that $W(s)$ is the inverse of $V(s)$, whereas $V[n]$ is a complex number. PV and PQ buses can be assessed using these equations. In PV buses, the voltage magnitude constraint is handled in (7.13) where $V_{\text{lre}}[n]$ corresponds to the *n*th coefficient in the voltage power series. This equation is used to calculate the unknown coefficients in the reactive power series and voltage of PVs and PQs by separating the imaginary and real parts of the equations and reorganizing the well-known terms:

$$
\begin{bmatrix}
-1 & 0 & 0 & 0 & 0 & 0 \\
0 & -1 & 0 & 0 & 0 & 0 \\
-G_{l1} & B_{l1} & 0 & B_{l2} & -G_{lN} & B_{lN} \\
-B_{l1} & -G_{l1} & -1 & -G_{l2} & -B_{lN} & -G_{lN} \\
-G_{j1} & B_{j1} & 0 & B_{j2} & -G_{jN} & B_{jN} \\
-B_{j1} & -G_{j1} & 0 & -G_{j2} & -B_{jN} & -G_{jN}
\end{bmatrix}
\begin{bmatrix}
V_{s_l \, re}[n] \\
V_{s_l \, im}[n] \\
Q_l[n] \\
V_{l \, im}[n] \\
V_{j \, re}[n] \\
V_{j \, im}[n]
\end{bmatrix}
$$

$$
=
\begin{bmatrix}
0 \\
0 \\
G_{l2} \\
B_{l2} \\
G_{j2} \\
B_{j2}
\end{bmatrix}
V_{l \, re}[n] +
\begin{bmatrix}
\delta_{n1}(1 - V_{slack}) - \delta_{n0} \\
0 \\
-\mathrm{Re}(Rh_l[n-1]) \\
-\mathrm{Im}(Rh_l[n-1]) \\
\mathrm{Re}(Y_{j \, sh} V_j[n-1] - S_j^* W_j^*[n-1]) \\
\mathrm{Im}(Y_{j \, sh} V_j[n-1] - S_j^* W_j^*[n-1])
\end{bmatrix}
-
\tag{7.14}
$$

In any system configuration with known PV buses, the LHS matrix is a constant. Our mismatch vector is calculated by evaluating the active power on generator buses and the complex power on load buses. The method is repeated until convergence has been achieved. As a new starting point for subsequent restarts, this method estimates the voltage at any point of the restart. HELF's main advantages are its non-iterative nature and the independence of the HELF solution from an initial guess, which is calculated using a constant matrix of iterations. Therefore, this method is better suited to calculating ATC accurately and quickly.

7.2.8   APEBS method

For a two-bus simple system, (7.15) represents the load flow equation. Equation (7.16) represents PEBS, and (7.17) represents the energy function [34]:

$$
\begin{aligned}
f_1(\delta, v) &= 0 = B_{12} v \sin(\delta) - P_L \\
f_2(\delta, v) &= 0 = B_{12} v \cos(\delta) + B_{22} v^2 - Q_L
\end{aligned}
\tag{7.15}
$$

$$
PEBS(\delta, v) = \frac{f_2(\delta, v)}{v}(v - v^s) + f_1(\delta, v)(\delta - \delta^s)
\tag{7.16}
$$

$$
V(\delta, v) = Q_L \ln(v) + P_L \delta - 0.5 B_{22} v^2 - B_{12} v \cos(\delta)
\tag{7.17}
$$

$$
\frac{\partial V}{\partial \delta} = f_1(\delta, v), \quad \frac{\partial V}{\partial v} = \frac{f_2(\delta, v)}{v}
\tag{7.18}
$$

The energy method is a more accurate approximation of PEBS than (7.19), known as APEBS

$$
APEBS(\delta, v) = \frac{\partial V}{\partial v}(v - v^s)e^{-v} + \frac{\partial V}{\partial \delta}(\delta - \delta^s)
\tag{7.19}
$$

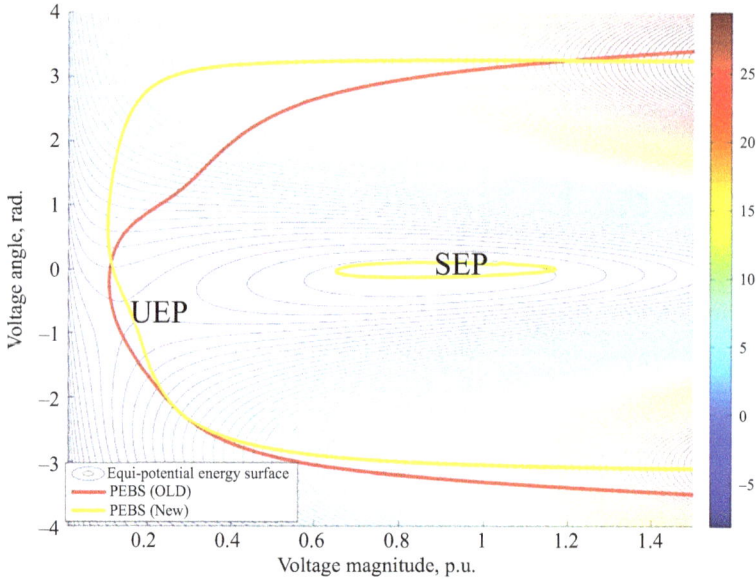

Figure 7.3   Comparison of PEBS and APEBS

For a two-bus system, Figure 7.3 illustrates the difference between APEBS and PEBS. In Figure 7.3, APEBS (yellow thick line), PEBS (red thick line), and the equipotential energy surface (thin line) can be seen.

7.2.9   Conclusion

DATC can still be determined using novel methods developed by improving and combining old methods. According to the simulation results, the following was found:

- The proposed method to evaluate DATC consumes much less CPU time than all repeated methods.
- There is relatively good agreement between the results of repeated methods and those of DATC, and DATC is highly accurate.
- Four improved methods are combined in the new approach.
- This method has produced acceptable results for a large real network that contains two standard networks and a large number of wind farms.
- This approach is optimal for both operation and planning.

7.3   DATC and DELF

ATC calculations are conducted using DELF rather than the usual load flow. Also, static ATC [67] can use the minimum distance path approximation method (AMD method) [26,27]. Instead of repeating the power flow, this method only uses three outputs for the load flow. In DELF, fictitious dynamic systems are created from load flow models. In DELF, the time-domain trajectory of a dynamic equation is analyzed rather than the power flow equations are repeatedly solved under the same conditions.

Additionally, PEBS estimation was utilized to correct previous estimations of transient stability. ATC computations are based on a combination of transient stability and static methods. It uses DELF and considers intact system and first contingency $(n-1)$ situations conditions to assess DATC. ATC dynamic methods were found to be more accurate, faster, and more suitable than other dynamic methods.

7.3.1   SATC and DELF

The DELF algorithm solves a linear equation only once for every phase of time, and a high-order approximation is used to reduce calculation time. In Section 7.3.2, the DELF procedure is shown to be effective for determining saddle-node bifurcations in power systems.

The proposed approach begins by evaluating the LF solution using DELF with $\lambda = 0$ at the power contract between sellers and buyers. As a result, we get three outputs for the load flow by a slightly increasing or decreasing loading parameter λ. Section 7.3.3 calculates the static ATC from these three points. ATC is estimated using this method at a reasonable speed according to [43]. In this method, AMD uses because accuracy does not matter, but speed does, therefore, AMD uses SATC.

As an upper bound for DATC, DELF is suggested to be used as the main method in the algorithm.

1. In Section 7.3.2, the (b) vector changes with every transaction between sellers and buyers.
2. Utilize AMD and DELF to find the maximum λ (Section 7.3.3) for this transaction.
3. ATC estimates the effect of generator and line outages separately.
4. Utilize AMD and DELF to find the maximum λ for each contingency (Step 3).
5. SATC is evaluated at each transaction by Min (λ_{max}).

In this method, the algorithm is exactly the same as that in Figure 7.1, Section 7.2.1, with the exception that method (HELF) is used instead of (DELF). The new approach is both fast and accurate. Using the energy method as an initial condition, the transient stability limit is determined by a voltage stability restriction (Section 7.2.8).

This study concluded that incorporating novel ideas and goals into the platform would enable the novel method to be used in actual network computation. The proposed method of evaluating DATC consumes considerably less CPU time than all repeated methods combined. A reasonable match can be made between the results of the simulation method and the DATC values obtained as a result.

7.3.2   *DELF*

The DELF approach, discussed in [35,58], is briefly outlined below. A typical load flow is illustrated in (7.20), where Б denotes a load factor direction, $\hat{S}$ is the complex power vector, $\tilde{V}$ is the bus voltage vector, $y=[e,f]$ is the rectangular matrix of the bus voltage vector, and g is the nonlinear field of the voltage vector on the bus:

$$\hat{S} = \tilde{V}\left(Ybus.\tilde{V}\right)^*, \hat{S} + \lambda Б = \tilde{V}.Ybus^*.\tilde{V}^* \rightarrow g(y,\lambda) = 0 \qquad (7.20)$$

Eq. (7.21) shows the function of a polynomial $y(t)$ and the coefficient of a power series of kth order $Y(k)$ [32–34]:

$$Y(k) = \frac{d^k y(t)}{dt^k}(k!)^{-1} \leftrightarrow y(t) = \sum_{k=0}^{K} Y(k).t^k \qquad (7.21)$$

The following rules must be followed to convert $x(t)$ to $X(k)$:

$$cx(t) \leftrightarrow cX(k), \ y(t) \pm x(t) \leftrightarrow Y(k) \pm X(k)$$
$$\dot{y} \leftrightarrow (k+1)Y(k+1)$$
$$c \leftrightarrow c\delta(k), \text{such that}: \delta(k) = \begin{cases} 0, k \neq 0 \\ 1, k = 0 \end{cases} \qquad (7.22)$$
$$y(t).x(t) \leftrightarrow \sum_{i=0}^{k} Y(i).X(k-i)$$

Below is a brief description of each step of the DELF method.

Step 1: Various differential-algebraic equations can be constructed by extending the algebraic equation (7.21) [35,58]:

$$g(y(t), \lambda(t)) = 0, \ h(x(t), y(t), \lambda(t)) = 0, \ f(x(t), y(t), \lambda(t)) = \dot{x}(t) \qquad (7.23)$$

Load flow equation (7.21) is dynamic as follows, with $v_l(t)$ representing the voltage in bus l and Ci denoting constant current.

$$\text{Form } I : \dot{\lambda}(t) = C_1, \ g(y(t), \lambda(t)) = 0 \qquad (7.24)$$

$$\text{Form } II : \dot{v}_l(t) = C_2, \ f_l^2(t) + e_l^2(t) - v_l^2(t) = 0, \ g(y(t), \lambda(t)) = 0 \qquad (7.25)$$

Formulas 1 and 2 calculate equilibrium points as well as maximum loading points.

Step 2: The following equations can be obtained from (7.22).

$$G(Y(0:k), \Lambda(0:k)) = 0$$
$$H(X(0:k), Y(0:k), \Lambda(0:k)) = 0 \qquad (7.26)$$
$$F(X(0:k), Y(0:k), \Lambda(0:k)) = (k+1)X(k+1)$$

Step 3: B is a function of $Y(0:k-1)$ and $\Lambda(0:k-1)$, whereas matrix A is a function of $Y(0)$ and $\Lambda(0)$. In the bus voltage vector, $Y(k)$ is the coefficient of a power series of kth order determined analytically from $\Lambda(0:k-1)$ and $Y(0:k-1)$. This depends on whether the auxiliary function h is necessary when designing a differential equation as in (7.23). References [35,58] provide more details:

$$A_{gy}Y(k) + A_{g\lambda}\Lambda(k) + B_g = 0, \; A_{hy}Y(k) + A_{h\lambda}\Lambda(k) + B_h = 0 \tag{7.27}$$

$$Y(k) = -A_{gy}^{-1}(B_g + A_{g\lambda}\Lambda(k)) \tag{7.28}$$

$$\begin{bmatrix} B_g \\ B_h \end{bmatrix} = -\begin{bmatrix} A_{gy} & A_{g\lambda} \\ A_{hy} & A_{h\lambda} \end{bmatrix}\begin{bmatrix} Y(k) \\ \Lambda(k) \end{bmatrix} \tag{7.29}$$

In formally linear equations (7.27) with matrices A and B, the differential transformation (7.26) of a nonlinear LF equation estimates a nonlinear load flow equation:

$$B_g = [B_{PQ} \; B_{PV} \; B_{REF}]^T, A_{g\lambda} = [A_{\lambda,PQ} \; A_{\lambda,PV} \; A_{\lambda,REF}]^T, A_{gy} = [A_{y,PQ} \; A_{y,PV} \; A_{y,REF}]^T,$$
$$A_{y,PQ} = [a_{P,1} \; a_{Q,1} \; \cdots \; a_{P,M} \; a_{Q,M}]^T, A_{\lambda,PQ} = [\Delta p_1 \; \Delta q_1 \; \cdots \; \Delta p_M \; \Delta q_M]^T$$
$$B_{PV} = [\varepsilon_{M+1} \; \mu_{M+1} \; \cdots \; \varepsilon_{N-1} \; \mu_{N-1}]^T, A_{y,PV} = [a_{P,M+1} \; a_{V,M+1} \; \cdots \; a_{P,N-1} \; a_{V,N-1}]^T$$
$$A_{\lambda,PV} = [\Delta p_{M+1} \; 0 \; \cdots \; \Delta p_{N-1} \; 0]^T, B_{PQ} = [\varepsilon_1 \; \mu_1 \; \cdots \; \varepsilon_M \; \mu_M]^T,$$
$$B_{REF} = [-e_N^{sp}\delta(k) \; -f_N^{sp}\delta(k)]^T, A_{y,REF} = [a_{E,N} \; a_{F,N}]^T, A_{\lambda,REF} = [0 \; 0]^T \tag{7.30}$$

Step 4: The algorithm developed a non-iterative algorithm to solve $t(k)$, $y(k)$, and $x(k)$ by using (7.26) and (7.29), and estimate variables $x(t)$, $y(t)$. It is possible to directly obtain the points of the LF equations after solving $y(t)$ and $\lambda(t)$:

$$x(t) = \sum_{i=0}^{K} t^i.X(i), \; y(t) = \sum_{i=0}^{K} t^i.Y(i), \; \lambda(t) = \sum_{i=0}^{K} t^i.\Lambda(i) \tag{7.31}$$

A basic idea used in Steps 1–4 of the algorithm is to solve $Y(k)$, $\Lambda(k)$, and $X(k)$. More information can be found in Ref. [35,58].

7.3.3 AMD method

This section discusses the application of the minimum distance (AMD) approach to approximating the path. (Min V), (Max V), and (Max loading) are the three constraints used by the AMD method to calculate SATC.

Due to the AMD approach, the operating point is defined under the constraints above:

$$MD_1 = (V_i - V_{\min})K_1 \quad i = 1:n \tag{7.32}$$

$$MD_2 = (V_{\max} - V_i)K_2 \quad i = 1:n \tag{7.33}$$

$$MD_3 = (L_{\max} - L_j)K_3 \quad j = 1:m \tag{7.34}$$

$$MDC = \text{Min}(MD_{1i}, MD_{2i}, MD_{3j}) \; \forall i,j \tag{7.35}$$

This research uses coefficients ($K_1 = K_2 = 1{,}000$, and $K_3 = 1$) to set loading and voltage. MDC (7.35) equals zero or close to zero when each constraint is met for each element. AMD is a quadratic approximation for modeling MDC. The quadratic function can be determined using three outputs of the load flow. The voltage

(VI) and overload (PI) criticality indexes are defined as follows:

$$VI_k = \sum_{i=1}^{Np} K_{Vi} \frac{(V_i^k - V_i^0)^2}{(V_i^0)^4} \left(\frac{P_{Load}^i}{P_{Load}^{max}}\right), \quad PI_k = \sum_{i=1}^{NL} K_{Pi} \left(\frac{S_i^k - S_i^o}{S_i^{max}}\right)^2 \left(\frac{S_i^o}{S_{max}}\right)$$

$$(7.36)$$

Where

$$K_{Vi} = \begin{cases} 1 & for \quad 1.05^{p.u} > V_i^o > V_i^k > 0.95^{p.u} \\ 10 & for \quad V_i^o > 0.95^{p.u}, V_i^k < 0.95^{p.u} \\ 5 & for \quad V_i^k < V_i^o < 0.95^{p.u} \end{cases}, \quad K_{Pi} = \begin{cases} 6 & if \quad S_i^o < S_i^k < S_i^{max} \\ 57 & if \quad S_i^o < S_i^k, S_i^k > S^{max} \\ 30 & if \quad S_i^k > S_i^o > S_i^{max} \end{cases}$$

$$(7.37)$$

where PI_k and VI_k are two critical indices of overload and voltage on the kth line, respectively, P_i^{Load} represents the load power in the ith bus, $P_{maxLoad}$ is the load power maximum, V_i^k is the ith stable bus voltage after the kth line is out of service, V_i^0 is the ith stable bus voltage before the kth line is out of service, S_i^{max} is the maximum capacity of the kth line, S_k^i is complex power the ith line after losing the kth line, S_i^0 is complex power the ith line before losing the kth line, and PQ buses are numbered N_p.

7.4   DATC and HVDC and wind

In this section, the backbone of the future transmission network is demonstrated as ultra-high voltage direct current lines, which are combined with Newton–Raphson–Seidel alternating LF equations. The proposed method uses matrix methods to accelerate load flow equation solving by increasing the dimensions of the load flow equations. Likewise, Newton–Raphson–Seidel load flow can be determined using this method since holomorphic load flow does not require an initial guess. The method presented in this section combines these methods to calculate the maximum static transmission power in small and large networks with appropriate accuracy and speed [61].

7.4.1   Importance of HVDC network

China has been trying to convince the world that it is the backbone of a "Global Energy Internet" since Chinese President Xi Jinping proposed it at the United Nations seven years ago. Essentially, it means building a line that carries ultra-high voltage direct current. But so far, Beijing has not seen any results from its plan to expand this intercontinental network. Nonetheless, super grids seem to be the future of power grids. The power network cannot be solved without the HVDC network in the future [68].

The most recent book in this field contains basic information on HVDC line modeling in the sixth chapter of the reference [69]. A number of inverter control models, frequency control models, and voltage-dependent load models are discussed in Ref. [70]. A network operation and control algorithm involving the connection of several control centers is also proposed in Ref. [71]. HVDC lines are modeled in this section using Ref. [62].

It is concluded from examining different methods of ATC calculation that, in the future, these methods will not only need to be more accurate and faster but also use different models of HVDC lines.

7.4.2   Mathematical model of AC/DC network

Figure 7.4 shows an HVDC line with a low-frequency filter, high-frequency filter, shunt capacitor, smoothing reactor, shunt capacitor DC, rectifier, and inverter [61].

In the AC network, the equations are as follows:

$$F(x) = 0 = \begin{bmatrix} P - f_P \\ Q - f_Q \end{bmatrix} \tag{7.38}$$

where

$$S_i = S_{Gi} - S_{Di} = P_{Gi} - P_{Di} + j(Q_{Gi} - Q_{Di}) = P_i + jQ_i \tag{7.39}$$

$$f_{Pi}(V, \delta) = V_i^2 G_{ii} + V_i \sum_{j=1 \neq i}^{n} V_j(G_{ij} \cos(\delta_i - \delta_j) + B_{ij} \sin(\delta_i - \delta_j)) \tag{7.40}$$

$$f_{Qi}(V, \delta) = -V_i^2 B_{ii} + V_i \sum_{j=1 \neq i}^{n} V_j(G_{ij} \sin(\delta_i - \delta_j) - B_{ij} \cos(\delta_i - \delta_j)) \tag{7.41}$$

An HVDC line connected between buses i and j changes (7.38) to (7.42) [71]:

$$\begin{aligned} \Delta P_i &= P_i - f_{Pi} - V_{dk}I_{dk}\mathrm{sign}(i) = 0 \\ \Delta Q_i &= Q_i - f_{Qi} - V_{dk}I_{dk}\tan(\varphi_k) = 0 \end{aligned} \tag{7.42}$$

where (V_{dk}), (I_{dk}), and (φ_k) are the voltage, current, and power factor angle of the DC line. The DC network equations can now be summarized as (7.43) [61,71]:

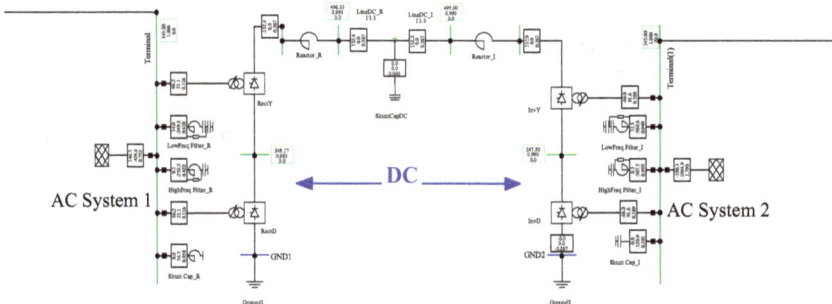

Figure 7.4   A HVDC line [61]

$$\begin{cases} \Delta d_{1k} = V_{dk} - k_{Tk}V_i \cos\theta_{dk} + X_{ck}I_{dk} \\ \Delta d_{2k} = V_{dk} - k_\gamma k_{Tk}V_i \cos\varphi_k \\ \Delta d_{3k} = sign(k)I_{dk} - \sum_{j=1}^{n_c} g_{dkj}V_{dj}, \qquad (k = 1 : n_c) \\ \Delta d_{4k} = d_{4k}(I_{dk}, V_{dk}, \cos\theta_{dk}, k_{Tk}) \\ \Delta d_{5k} = d_{5k}(I_{dk}, V_{dk}, \cos\theta_{dk}, k_{Tk}) \end{cases} \qquad (7.43)$$

If $X_{AC}=[\delta,V]$ represents AC network variables, and $X_{DC}=[V_{dk},I_{dk},\theta_{dk},k_{Tk},\varphi_k]$ represents DC network variables, and the main variables of AC load flow equations are similar to DC load flow equations, the AC/DC network load flow equations can be modeled as (7.44):

$$\overbrace{\begin{bmatrix} \Delta P \\ \Delta Q \\ \Delta d \end{bmatrix}}^{\Delta F} = \overbrace{\begin{bmatrix} J_{P\theta} & J_{PV} & J_{PX_{DC}} \\ J_{Q\theta} & J_{QV} & J_{QX_{DC}} \\ J_{d\theta} & J_{dV} & J_{dX_{DC}} \end{bmatrix}}^{J} \overbrace{\begin{bmatrix} \Delta\theta \\ \Delta V \\ \Delta X_{DC} \end{bmatrix}}^{\Delta X} \qquad (7.44)$$

In this section, (7.44) is solved using the Newton–Raphson–Seidel (NRS) method, which will be covered in the following section on static ATC calculation.

7.4.3 Solving the AC/DC load flow equation

NRS equations are given in Section 7.2.5 [28,57], and they are modified for AC/DC networks in this section:

$$\Xi(\xi) = \begin{bmatrix} (J - \lambda * I) * v \\ \|v\| - 1 \\ F(X) - y_0 - p * \alpha \end{bmatrix}, \xi = \begin{bmatrix} X \\ v \\ \alpha \end{bmatrix} \qquad (7.45)$$

Equation (7.45) requires its derivative to be solved:

$$\Xi_\xi(\xi) = \frac{\partial\Xi(\xi)}{\partial\xi} = \begin{bmatrix} \Xi_{\xi 11} & \Xi_{\xi 12} \\ \Xi_{\xi 21} & \Xi_{\xi 22} \end{bmatrix} \qquad (7.46)$$

$$\Xi_{\xi 11} = \frac{\partial^2 F}{\partial X^2} * diag(v_1, v_2, ..., v_n) \qquad (7.47)$$

$$\Xi_{\xi 12} = \begin{bmatrix} J - \lambda * I & 0 \\ \dfrac{v'}{\|v\|} & 0 \end{bmatrix}, \Xi_{\xi 21} = \frac{\partial F}{\partial X}, \Xi_{\xi 22} = \begin{bmatrix} 0 & -p \end{bmatrix} \qquad (7.48)$$

A Down-Hill method is used to speed up convergence (Section 7.2.6):

$$\xi^{(r+1)} = \xi^{(r)} - D_h \times inv\left(\left[\Xi_\xi(\xi^{(r)})\right]\right)\Xi(\xi^{(r)}) \qquad (7.49)$$

This inverse matrix can be used to increase speed, similar to the one described in Section 7.2.6:

$$inv([\Xi_\xi(\xi)]) = \begin{bmatrix} \Xi_{\xi11}^{-1} + \Xi_{\xi11}^{-1}\Xi_{\xi12}\Xi_Z\Xi_{\xi21}\Xi_{\xi11}^{-1} & -\Xi_{\xi11}^{-1}\Xi_{\xi12}\Xi_Z \\ -\Xi_Z\Xi_{\xi21}\Xi_{\xi11}^{-1} & \Xi_Z \end{bmatrix},$$ (7.50)

$$\Xi_Z = inv(\Xi_{\xi22} - \Xi_{\xi21}\Xi_{\xi11}^{-1}\Xi_{\xi12})$$

7.4.4 SATC and holomorphic method

This section uses the holomorphic algorithm from the third chapter.

1. Enter complete network information
2. Choosing a contract, determining the seller and buyer
3. Select an event from among $n-1$ events, line exit, generator, or transformer
4. Run HEPF load flow to determine the appropriate starting point for load flow
5. Implementation of NRS load distribution with and without HVDC line and determination of static ATC in this case
6. If all events have been considered, move on to step 6, otherwise step 3 should be repeated.
7. If all contracts have been checked, the next step can be taken, otherwise, step 2 should be repeated.

7.4.5 Conclusion

It is inevitable that HVDC lines will be added to all transmission networks in the near future, so it is imperative that power transmission calculation methods can analyze these transmission networks even though they are complex. Due to combining the DC network equations with the Newton–Raphson–Seidel load flow equations, the Jacobian matrix in the current section has grown in size. Consequently, the inverse Jacobian matrix developed increased the speed of network load flow calculations. Then, to get a better start, holomorphic load distribution was used. By selecting a suitable point of entry, Newton–Raphson–Seidel load flows were accelerated. After this method is applied to the SATC of this section, it is necessary to apply it to the dynamic ATC. This should be calculated by combining the methods used for determining transient stability with the ones used for determining static ATC.

7.5 ATC and state estimation

A combined method for dynamic ATC based on online state estimation (ES), static methods, and transient stability is discussed in the present section [59,60]. Power system computations have been influenced by renewable energy and are moving towards probabilistic methods rather than deterministic methods. Instead of the usual load flow, probabilistic load flow (PLF) is recommended [72–74]. The LF parameters should also be determined using the state estimation (ES) program [59,60]. A summary of the benefits of state estimation is provided in [59,60]: correction of measurement errors, estimation of unmeasured quantities, detection of inappropriate measurements, and, consequently, increasing LF accuracy.

Programming by DIgSILENT is used in conjunction with software developed by MatrikonOPC and DIgSILENT. Figure 7.5 [59,60] summarizes the algorithm used in the software.

The present section contains measurements taken out of the circuit, some of which have noise and send incorrect values. Due to the incomplete information, LF cannot be obtained, and only the state ES program is capable of detecting the incorrect information and obtaining LF in this scenario. Figure 7.6 illustrates the online state ES algorithm in DIgSILENT software with Matrikon software.

In this chapter, ES and wind farms are performed in a real network to determine the online DATC. In this algorithm, the following two sections are added instead of other sections in other algorithms:

1. It is not feasible for LF implementation to use consumer information, and the network is considered invisible.
2. On DIgSILENT, a completed state estimation program run on the Matrikon server simulator detects incorrect information about power flow.

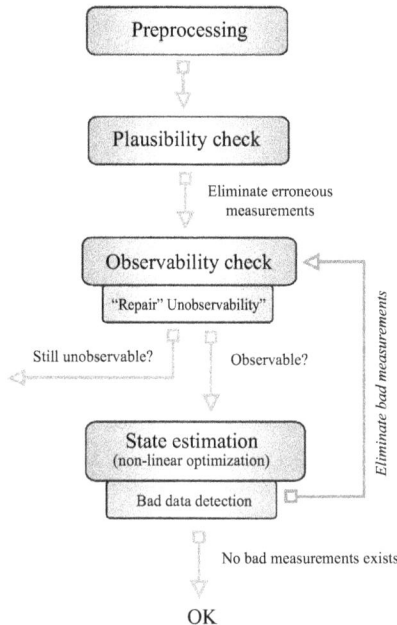

Figure 7.5   The DIgSILENT state estimation algorithm

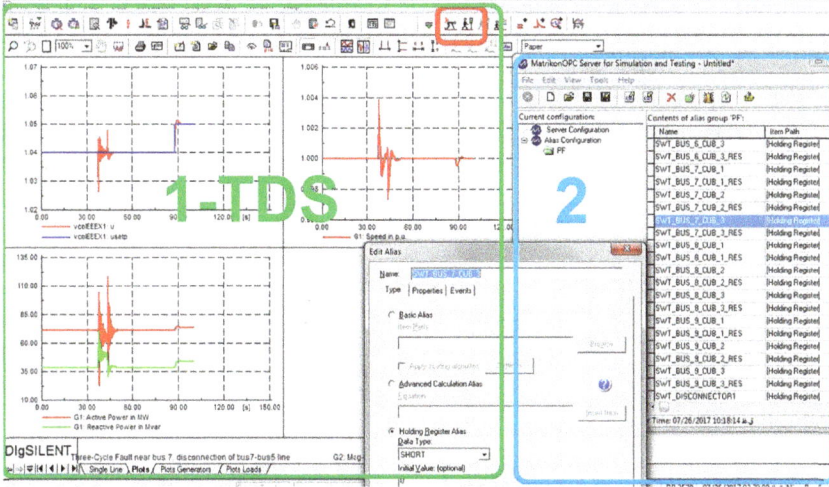

Figure 7.6 Online state ES algorithm in DIgSILENT software with Matrikon software

As a final note, the state estimation program is mandatory if there is incomplete network information. It is inevitable to use industrial software instead of academic software in real networks.

7.6 ATC and cyber security

Cyberattacks are simulated in DIgSILENT for state estimation and energy management in this section. In the event, hackers are able to manipulate input data, they could alter the ATC and power market. False information coordinated with error information may result in detection programs failing. State estimation requires 55% of the total measurements before it converges. The novel method accelerates state estimation by using weighted least squares. Additionally, the novel method improves ATC's speed. This algorithm can be used to identify the target of a cyberattack. In addition to correct and complete information, state estimator errors can also be reported through incorrect measurements. There is an inverse relationship between ATC values and loss and load [4].

7.6.1 Power system cybersecurity

Digitalization of power systems has allowed them to be controlled online and digitally connected [75]. SCADA systems use legacy ICT which makes them an attractive target for adversaries [76]. Network security is not built into most SCADA protocols [77]. Cyberattacks can cause cascading outages and even affect the economic aspects of the power system [78]. A detailed description of the risks posed by adversary attacks and deliberate malware can be found in [79]. As a combination of advanced skills, in-depth knowledge of the system, impressive attack capabilities, and excellent stealth, the Stuxnet worm is one of the most dangerous cyber threats ever created [80].

A bad data detector (BDD) filters out possible false measurements caused by sensor defects or abnormalities in SCADA systems [81]. A number of statistical techniques have been discussed in [82], including sequential detection using the cumulative sum algorithm (CUSUM). Stealth attack detectors in [83] generate detection signals that include additional information like phasor measurement unit (PMU) data, generation schedules, and forecasted load amounts. Using an internet of things (IoT) platform and deep learning neural networks, [84] presents an innovative way of monitoring computers that are numerically controlled online. Based on a dynamic risk assessment model, [85] investigates cyber-physical power systems (CPPS). SCADA systems in substations can pose a network security vulnerability and cause physical damage to power systems when maliciously controlled.

The state estimation programs in ATC can be modified by cyberattacks. If the state estimation input is deliberately changed, the ATC output data will be incorrect. Due to this, the output error value may also be affected by the input error value.

Here is a summary of the implications of this section:

1. ATC information was changed by cyberattacks using the state estimation program in this study to demonstrate the significance of cyberattacks.
2. State estimation is speeded up using the updated WLS method.
3. It is necessary to use false measuring devices to detect hacking of state estimation information.
4. ATC is determined using both static and dynamic methods.

In most references, the model-based detector is improved by using a high-quality simulator [4].

7.6.2   *WLS method*

Transmission state estimation techniques have been developed and applied for more than 30 years. State estimation is most commonly performed using weighted least squares. State variables are correlated with measurements using a mathematical model. Measurements are contained in vector Z, errors in vector V, and non-linear measurement functions are contained in vector h(x) [4]:

$$Z = V + h(x) \qquad (7.51)$$

This optimization problem [4] is used to estimate the WLS state of the system.

$$\text{Min } J(x) = \sum_{i=1}^{m} w_i (z_i - h_i(x))^2 = [Z - h(x)]^T W [Z - h(x)] \qquad (7.52)$$

Weights can be found from w_i, but residual vectors or matrices can be found from $r = Z - h(x)$. Measurement weight coefficients are divided into diagonal elements equal to the inverse measurement variance. $h_i(x)$ represents the measurement function proportional to the measurements Z_i has taken [4].

The objective function in the next equation should be minimized using state estimation vectors [4], and the Jacobian matrix is $H(x)$:

$$g(x) = \frac{\partial J(x)}{\partial x} = H^T(x).W.[Z - h(x)] = 0, \; H(x) = \frac{\partial h(x)}{\partial x} \tag{7.53}$$

This equation can be obtained by employing Newton's iterative method:

$$x^{k+1} = x^k - G(x)^{-1}H^T(x^k).W.[h(x^k) - Z]$$
$$s.t. : \; H^T(x^k).W.H(x^k) = G(x) \tag{7.54}$$

The two G and H matrices are defined by the expression $x = (\delta, V)$ [4]:

$$G = \begin{bmatrix} G_{11} & G_{12} \\ G_{21} & G_{22} \end{bmatrix}, H = \begin{bmatrix} H_{11} & H_{12} \\ H_{21} & H_{22} \end{bmatrix} \tag{7.55}$$

where

$$G = \begin{bmatrix} G_{11} & G_{12} \\ G_{21} & G_{22} \end{bmatrix} = \begin{bmatrix} H_{11}^T & H_{12}^T \\ H_{21}^T & H_{22}^T \end{bmatrix}.\begin{bmatrix} W_1 & 0 \\ 0 & W_2 \end{bmatrix}.\begin{bmatrix} H_{11} & H_{12} \\ H_{21} & H_{22} \end{bmatrix}$$

$$G = \begin{bmatrix} H_{12}^T.W_2.H_{12} + H_{11}^T.W_1.H_{11} & H_{12}^T.W_2.H_{22} + H_{11}^T.W_1.H_{12} \\ H_{22}^T.W_2.H_{12} + H_{12}^T.W_1.H_{11} & H_{22}^T.W_2.H_{22} + H_{12}^T.W_1.H_{12} \end{bmatrix} \tag{7.56}$$

Matrix G's inverse is calculated as follows [35,57,61].

$$inv(G) = \begin{bmatrix} G_{11}^{-1} + G_{11}^{-1}G_{12}(G_{22} - G_{21}G_{11}^{-1}G_{12})^{-1}G_{21}G_{11}^{-1} & -G_{11}^{-1}G_{12}(G_{22} - G_{21}G_{11}^{-1}G_{12})^{-1} \\ -(G_{22} - G_{21}G_{11}^{-1}G_{12})^{-1}G_{21}G_{11}^{-1} & (G_{22} - G_{21}G_{11}^{-1}G_{12})^{-1} \end{bmatrix} \tag{7.57}$$

A simplified version of the equations can be obtained by relating $(Q-V)$ and $(P-\delta)$ [4].

$$\begin{cases} H_{12} = 0 \\ H_{21} = 0 \end{cases} \Rightarrow G = \begin{bmatrix} H_{11}^T.W_1.H_{11} & 0 \\ 0 & H_{22}^T.W_{Q2}.H_{22} \end{bmatrix} \Rightarrow G^{-1} = \begin{bmatrix} G_{11}^{-1} & 0 \\ 0 & G_{22}^{-1} \end{bmatrix} \tag{7.58}$$

This method determines ATC using the following algorithm.

7.6.3   The suggested algorithm

The main algorithm in this section is summarized below [4]:

1. Information about the entire network is received.
2. In the network, power buyers and sellers are identified.

3. Fast-state estimation is used to estimate measurement values and false measurements.
    3.1 A state estimation database can store the location and number of false measuring devices.
    3.2 The state estimation information hacking warning appears if the measurement information is accurate and no incorrect measurements were taken.
4. Using the DELF method [25], the SATC is the upper bound of the DATC.
5. DATC-DE calculates dynamic ATC [35].
6. The algorithm continues in Step 2 until all sellers and buyers are exhausted.

In the above algorithm, the following points are taken into account. Detection algorithms that take advantage of the natural weaknesses of networks seldom produce false positives or inaccurate measurements. It is assumed that all network information was compromised. The hacker hacked the entire state estimation program, detecting all measurement data as accurate.

7.6.4   With/without cyberattacks in ATC

There were 24 measurements throughout the 24-h simulation. ATC values have been changed by two cyberattacks in the last 24 h to disrupt the electricity market. The proposed program and algorithm in this section can be used to identify attackers' targets as well as to detect malicious cyberattacks, as can be seen in Figure 7.7.

Different contracts have proportional ATCs, loads, and losses. There is a certain pattern to the values, even though they are not linear. It is assumed that the two cyberattacks at (7 a.m.) and (2 p.m.) changed the ATC values. Since there is complete information, state estimation is error-free. Other network parameters and ATC values are not in line with the circadian pattern at (7 a.m.) and (2 p.m.). The state estimation input data have been deliberately manipulated. As shown in Figure 7.7, it is important to compare current information with previous days or weeks. Loss and load are generally inversely related to ATC values.

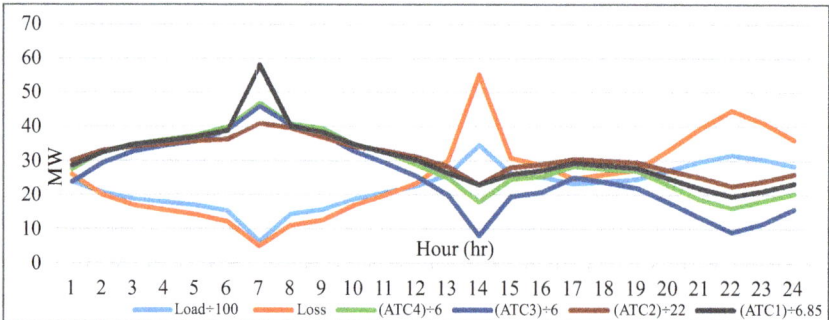

Figure 7.7   ATC, load, and loss after cyberattack

Hackers can obtain both complete and incomplete information, as shown in this section. State estimation program can be detected when incorrect measurements are used instead of complex methods. The security of multi-regional AGCs and LFCs must also be reviewed from a cyber perspective. In the future, researchers will continue to investigate attack points on AGC and LFC, as well as different attack models and attack strategies. Their task will be to evaluate the existing mechanisms for detecting and defending against such attacks. When LFC and AGC control systems are attacked, frequency or voltage can be destroyed.

7.7 Conclusion

This chapter of the book examines power system strength from ATC's perspective. To speed up calculations, static ATC determination methods simplify assumptions. Each assumption has a significant impact on ATC determination accuracy. Because static methods are fast, they are appropriate for online situations. However, they are not very accurate. It is an excellent idea to start with static ATC methods when determining dynamic ATC. In addition, novel algorithms are required to determine ATC in the future for all networks using renewable energy sources [86–89] and high-voltage direct current lines (HVDC). Power system calculations have become more complex due to wrong information, both normal and intentional (cyber-attack). The following are future works:

- ATC should be investigated in relation to stationary and mobile energy storage systems (ESS).
- Real networks should be used to implement the proposed algorithms.
- Future studies should be conducted for 2030 and beyond.
- Calculations should take into account the joint planning and operation of network resilience measures.
- It has been identified that cyberattacks on ATC will be the most common challenge in the future of cyber-physical power.
- The controllers need sufficient data storage to train the artificial intelligence needed to control and protect the system dynamically.
- FACTS tools remain focused on their location and optimal performance.

References

[1] M. Eidiani, H. Zeynal, and M. Shaaban, "A detailed study on prevailing ATC methods for optimal solution development," In *2022 IEEE International Conference on Power and Energy (PECon)*, 2022, pp. 299–303, doi:10.1109/PECon54459.2022.9988775.

[2] M. Eidiani and H. Zeynal, "New approach using structure-based modeling for the simulation of real power/frequency dynamics in deregulated power systems," *Turkish Journal of Electrical Engineering and Computer Sciences*, 2014, 22(5), pp. 1130–1146, https://doi.org/10.3906/elk-1208-90.

[3]   O. O. Mohammed, M. W. Mustafa, D. S. S. Mohammed, and A. O. Otuoze, "Available transfer capability calculation methods: a comprehensive review," *International Transactions on Electrical Energy Systems*, 2019, 2023, pp. 1–24.

[4]   M. Eidiani, "A rapid state estimation method for calculating transmission capacity despite cyber security concerns," *IET Generation, Transmission & Distribution*, 2023, pp. 1–9. https://doi.org/10.1049/gtd2.12747

[5]   M. E. Baydokhty, M. Eidiani, H. Zeynal, H. Torkamani, and H. Mortazavi, "Efficient generator tripping approach with minimum generation curtailment based on fuzzy system rotor angle prediction," *Przeglad Elektrotechniczny*, 2012, 88(9 A), pp. 266–271.

[6]   M. Eidiani, M. Ebrahimean Baydokhty, M. Ghamat, H. Zeynal, and H. Mortazavi, "Transient stability enhancement via hybrid technical approach," in: *2011 IEEE Student Conference on Research and Development*, 2011, pp. 375–380, doi:10.1109/SCOReD.2011.6148768.

[7]   M. Eidiani, M. E. Badokhty, M. Ghamat, and Zeynal, H., "Improving transient stability using combined generator tripping and braking resistor approach", *International Review on Modelling and Simulations*, 2011, 4(4), pp. 1690–1699.

[8]   A., Hassan Haes, M. M. Antonio, and S. Pierluigi (ed.), Industrial Demand Response: Methods, Best Practices, Case Studies, and Applications, IET Digital Library, Energy Engineering, 2022, doi:10.1049/PBPO215E.

[9]   V. G. Lopes and F. L. Lirio, "Power systems performance evaluation on high renewable penetration levels," *Sociedade Brasileira de automatica*, 2020, 1 (1), SBSE2020.

[10] L. Yu, K. Meng, W. Zhang, and Y. Zhang, "An overview of system strength challenges in Australia's National Electricity Market Grid," *Electronics*, 2022, 11, pp. 224. https://doi.org/10.3390/electronics11020224.

[11] P. C. S. Krishayya, R. Adapa, and M. Holm, IEEE Guide for Planning DC Links Terminating at AC Locations Having Low Short-Circuit Capacities; Part I: AC/DC System Interaction Phenomena. CIGRE: Paris, France, 1997, pp. 1–216.

[12] O. Damanik, Ö. C. Sakinci, G. Grdenić, and J. Beerten, "Evaluation of the use of short-circuit ratio as a system strength indicator in converter-dominated power systems," in: *2022 IEEE PES Innovative Smart Grid Technologies Conference Europe (ISGT-Europe)*, Novi Sad, Serbia, 2022, pp. 1–5, doi:10.1109/ISGT-Europe54678.2022.9960381.

[13] M. Maharjan, A. Ekic, and D. Wu, "Probabilistic grid strength assessment of power systems with uncertain renewable generation based on probabilistic collocation method," in: *2022 17th International Conference on Probabilistic Methods Applied to Power Systems (PMAPS)*, Manchester, UK, 2022, pp. 1–6, doi:10.1109/PMAPS53380.2022.9810616.

[14] N. Upadhayay, M. Nadarajah, and A. Ghosh, "System strength enhancement with synchronous condensers for power systems with high penetration of renewable energy generators," in: *2021 31st Australasian Universities*

Power Engineering Conference (AUPEC), Perth, Australia, 2021, pp. 1–5, doi:10.1109/AUPEC52110.2021.9597835.

[15] Summary results, by Region, of a recent transmission reliability margin (TRM) and capacity benefit margin (CBM) survey conducted by the ATCWG, ftp://ftp. nerc.com/pub/sys/all_updl/ac/atcwg/ svyrsult.pdf, 1999.

[16] M. Eidiani, M. H. M. Shanechi, and E. Vaahedi, "Fast and accurate method for computing FCTTC (first contingency total transfer capability)," in: *Proceedings. International Conference on Power System Technology*, 2002, 2, pp. 1213–1217, doi:10.1109/ICPST.2002.1047595.

[17] H. Zeynal, A. K. Zadeh, K. M. Nor, and M. Eidiani, "Locational Marginal Price (LMP) assessment using hybrid active and reactive cost minimization", *International Review of Electrical Engineering*, 2010, 5(5), pp. 2413–2418.

[18] V. Ajjarapu and B. Lee, "Bibliography on voltage stability," *IEEE Transactions on Power System*, 1998, 13(1), pp. 115–125.

[19] S. Grijalva and P. W. Sauer, "Transmission Loading Relief (TLR) and hour-ahead ATC," in: *Proceedings of the33rd Hawaii International Conference on System Sciences Copyright (c) IEEE*, 1998.

[20] G. C. Ejebe, J. Tong, J. Frame, X. Wang, and W. F. Tinney, "Available transfer capability calculations," Preprint Order Number: PE-321-PWRS-0-10-1997, Discussion Deadline: March 1998.

[21] G. Hamoud, "Assessment of available transfer capability of transmission systems," Preprint Order Number: PE-002PRS (09-99), Discussion Deadline: February 2000.

[22] M. Eidiani, "Assessment of voltage stability with new NRS," in: *2008 IEEE 2nd International Power and Energy Conference*, 2008, pp. 494–496, doi:10.1109/PECON.2008.4762525.

[23] M. Eidiani, Y. Ashkhane, and M. Khederzadeh, "Reactive power compensation in order to improve static voltage stability in a network with wind generation," in: *2009 International Conference on Computer and Electrical Engineering*, ICCEE 2009, 2009, 1, pp. 47–50, 5380672, doi:10.1109/ICCEE.2009.239.

[24] M. Eidiani, H. Zeynal, A. K. Zadeh, S. Mansoorzadeh, and K. M. Nor, "Voltage stability assessment: an approach with expanded Newton Raphson-Sydel," in: *2011 5th International Power Engineering and Optimization Conference*, 2011, pp. 31–35, doi:10.1109/PEOCO.2011.5970424.

[25] M. Eidiani, H. Zeynal, A. K. Zadeh, and K. M. Nor, "Exact and efficient approach in static assessment of Available Transfer Capability (ATC)," in: *2010 IEEE International Conference on Power and Energy*, 2010, pp. 189–194, doi: 10.1109/PECON.2010.5697580.

[26] M. Eidiani, S. M. Asadi, S. A. Faroji, M. H. Velayati, and D. Yazdanpanah, "Minimum distance, a quick and simple method of determining the static ATC," in: *2008 IEEE 2nd International Power and Energy Conference*, 2008, pp. 490–493, doi:10.1109/PECON.2008.4762524.

[27] M. Eidiani and D. Yazdanpanah, "Minimum distance, a quick and simple method of determining the static ATC", *Journal of Electrical Engineering*, 2011, 11(2), p. 16, pp. 95–101.

[28] M. Eidiani, "A reliable and efficient method for assessing voltage stability in transmission and distribution networks", *International Journal of Electrical Power and Energy Systems*, 2011, 33(3), pp. 453–456, doi:10.1016/j. ijepes.2010.10.007.

[29] M. Eidiani, "A new method for assessment of voltage stability in transmission and distribution networks", *International Review of Electrical Engineering*, 2010, 5(1), pp. 234–240.

[30] T. Nireekshana, G. K. Rao, and S. S. Raju, "Available transfer capability enhancement with FACTS using cat swarm optimization," *Ain Shams Engineering Journal*, 2016, 1, pp. 159–167.

[31] W. Xiaoting, W. Xiaozhe, H. Sheng, and X. Lin, "A data-driven sparse polynomial chaos expansion method to assess probabilistic total transfer capability for power systems with renewables," *IEEE Transactions on Power Systems*, Early Access, 28 October 2020.

[32] H. Zeynal, L. X. Hui, Y. Jiazhen, M. Eidiani, and B. Azzopardi, "Improving Lagrangian relaxation unit commitment with Cuckoo Search Algorithm," in: *2014 IEEE International Conference on Power and Energy (PECon)*, 2014, pp. 77–82, doi:10.1109/PECON.2014.7062417.

[33] T. L. Duong, T. T. Nguyen, N. A. Nguyen, and T. Kang, "Available transfer capability determination for the electricity market using Cuckoo Search Algorithm," *Engineering, Technology & Applied Science Research*, 2020, 10, pp. 5340–5345.

[34] M. Eidiani and M. H. M. Shanechi, "FAD-ATC: a new method for computing dynamic ATC", *International Journal of Electrical Power and Energy Systems*, 2006, 28(2), pp. 109–118, https://doi.org/10.1016/j.ijepes. 2005.11.004.

[35] M. Eidiani, "A new load flow method to assess the static available transfer capability," *Journal of Electrical Engineering and Technology*, 2022, 17(5), pp. 2693–2701, https://doi.org/10.1007/s42835-022-01105-3

[36] A. P. S. Meliopoulos, S. W. Kang, and G. Cokkinides, "Probabilistic transfer capability assessment in a deregulated environment," in: *Proceedings of the 33rd Hawaii International Conference on System Sciences Copyright (c) 1998 Institute of Electrical and Electronics Engineers*, 1998.

[37] M. H. Gravener, C. Nwankpa, and T. S. Yeoh, "ATC computational issues," in: *Proceeding of the 32nd Hawaii International Conference on System Science*, 1999.

[38] S. Grijalva and P. W. Sauer, "Reactive power considerations in linear ATC computation," in: *Proceedings of the 33rd Hawaii International Conference on System Sciences Copyright (c) IEEE*, 1998.

[39] Y. Zan, M. Shaaban, M. Li, H. Liu, Y. Ni, and F. Wu, "ATC calculation with steady-state security constraints using benders decomposition," *IEE*

Proceedings Generation and Transmission Distribution, 2003, 150(5), pp. 611–615.

[40]   L. Peijie, Z. Ling, B. Xiaoqing, and W. Hua, "Available transfer capability calculation constrained with small-signal stability based on adaptive gradient sampling," in: *Advanced Control and Optimization for Complex Energy Systems*, 2020.

[41]   S. Sayah and A. Hamouda, "Optimal power flow solution of integrated AC-DC power system using enhanced differential evolution algorithm," *International Transactions on Electrical Energy Systems*, 2018, 29, pp. e2737.

[42]   M. Eidiani, N. Asghari Shahdehi, and H. Zeynal, "Improving dynamic response of wind turbine driven DFIG with novel approach," in: *2011 IEEE Student Conference on Research and Development*, 2011, pp. 386–390, doi:10.1109/SCOReD.2011.6148770.

[43]   M. Eidiani, M. O. Buygi, and S. Ahmadi, "CTV, complex transient and voltage stability: A new method for computing dynamic ATC", *International Journal of Power and Energy Systems*, 2006, 26(3), pp. 296–304, https://doi.org/10.2316/Journal.203.2006.3.203-3597, 2006.

[44]   R. A. Schluter and A. Costi, "Multiple contingency selection for transmission reliable and transfer capability studies," *Electrical Machines and Power System*, 1992, 20(3), pp. 223–237.

[45]   A. L. Bettiol, L. Wehenkel, and M. Pavella, "Transient stability constrained maximum allowable transfer," *IEEE Transactions on Power System*, 1999, 14(2), pp. 654–659.

[46]   M. Eidiani, "Atc evaluation by CTSA and POMP, two new methods for direct analysis of transient stability," in: *IEEE/PES Transmission and Distribution Conference and Exhibition*, 2002, 3, pp. 1524–1529, doi:10.1109/TDC.2002.1176824.

[47]   Y. Xingbin and C. Singh, "Probabilistic analysis of total transfer capability considering security constraints," in: *8th International Conference on Probabilistic Methods Applied to Power Systems*, 2004, pp. 242–247.

[48]   J. Kubokawa, Y. Yuan, N. Yorino, Y. Zoka, Y. Sasaki, and L. Hakim, "A solution of total transfer capability using transient stability constrained optimal power flow," in: *IEEE Lausanne Power Tech*, 2007, pp. 2018–2022.

[49]   D. V. Kumar and C. Venkaiah, "Dynamic Available Transfer Capability (DATC) computation using intelligent techniques," in: *Joint International Conference on Power System Technology and IEEE Power India Conference, 2008. POWERCON 2008*, 2008, pp. 1–6.

[50]   C. Venkaiah, D. Kumar, and K. Murali, "Dynamic ATC computation for real-time power markets," *Journal of Electrical Engineering and Technology* 2010, 5(2), pp. 209–219.

[51]   A. Srinivasan, N. Ramkumar, B. Dineshkumar, and P. Venkatesh, "Support vector regression based dynamic available transfer capability in deregulated power market," in: *Circuits, Power and Computing Technologies (ICCPCT)*, 2013, pp. 68–73.

[52]   A. Srinivasan, P. Venkatesh, B. Dineshkumar, and N. Ramkumar, "Dynamic available transfer capability determination in power system restructuring environment using support vector regression," *International Journal of Electrical Power and Energy Systems,* 2015, 69, pp. 123–130.

[53]   A. I. S. Velusamy, N. B. Ramu, D. Durairaj, and K. Murugesan, "Differential evolutionary algorithm-based optimal support vector machine for online dynamic available transfer capability estimation incorporating transmission capacity margins," *International Transactions on Electrical Energy Systems,* 2017, 27(7).

[54]   M. Shaaban, "Behavior of power system equilibrium points in dynamic available transfer capability calculation," *Journal of Applied Science & Process Engineering,* 2018, 5(1), pp. 242–248.

[55]   M. Eidiani and H. Zeynal, "A fast holomorphic method to evaluate available transmission capacity with large scale wind turbines," in: *2022 9th Iranian Conference on Renewable Energy & Distributed Generation (ICREDG),* 2022, pp. 1–5, doi: 10.1109/ICREDG54199.2022.9804527.

[56]   M. Eidiani, H. Zeynal, and Z. Zakaria, "An efficient holomorphic based available transfer capability solution in presence of large scale wind farms," in: *2022 IEEE International Conference in Power Engineering Application (ICPEA),* 2022, pp. 1–5, doi:10.1109/ICPEA53519.2022.9744711.

[57]   M. Eidiani, "A reliable and efficient holomorphic approach to evaluate dynamic available transfer capability", *International Transactions on Electrical Energy Systems,* 2021, 31(11), e13031, pp. 1–14, https://doi.org/10.1002/2050-7038.13031, 2021.

[58]   M. Eidiani, "An efficient differential equation load flow method to assess dynamic available transfer capability with wind farms", in: *IET Renewable Power Generation,* 2021, pp. 3843–3855, https://doi.org/10.1049/rpg2.12299.

[59]   M. Eidiani and H. Zeynal, "Determination of online DATC with uncertainty and state estimation," in: *2022 9th Iranian Conference on Renewable Energy & Distributed Generation (ICREDG),* 2022, pp. 1–6, doi: 10.1109/ICREDG54199.2022.9804581.

[60]   M. Eidiani, H. Zeynal, and Z. Zakaria, "Development of online dynamic ATC calculation integrating state estimation," in: *2022 IEEE International Conference in Power Engineering Application (ICPEA),* 2022, pp. 1–5, doi:10.1109/ICPEA53519.2022.9744694.

[61]   M. Eidiani, "A new hybrid method to assess available transfer capability in AC–DC networks using the wind power plant interconnection," *IEEE Systems Journal,* 2022, 17(1), pp. 1375–1382, doi:10.1109/JSYST. 2022.3181099.

[62]   H. Zeynal, Y. Jiazhen, B. Azzopardi, and M. Eidiani, "Flexible economic load dispatch integrating electric vehicles," in: *2014 IEEE 8th International Power Engineering and Optimization Conference (PEOCO2014),* 2014, pp. 520–525, doi:10.1109/PEOCO.2014.6814484.

[63]   H. Zeynal, Y. Jiazhen, B. Azzopardi, and M. Eidiani, "Impact of Electric Vehicle's integration into the economic VAr dispatch algorithm," in: *2014*

IEEE Innovative Smart Grid Technologies – Asia (ISGT ASIA), 2014, pp. 780–785, doi:10.1109/ISGT-Asia.2014.6873892.

[64] M. Eidiani and A. Ghavami, "New network design for simultaneous use of electric vehicles, photovoltaic generators, wind farms and energy storage," in: *2022 9th Iranian Conference on Renewable Energy & Distributed Generation (ICREDG)*, 2022, pp. 1–5, doi: 10.1109/ICREDG54199.2022.9804534.

[65] M. Eidiani, H. Zeynal, A. Ghavami, and Z. Zakaria, "Comparative analysis of mono-facial and bifacial photovoltaic modules for practical grid-connected solar power plant using PVsyst," in: *2022 IEEE International Conference on Power and Energy (PECon)*, 2022, pp. 499–504, doi:10.1109/PECon54459.2022.9988872.

[66] H. Zeynal and M. Eidiani, "Hydrothermal scheduling flexibility enhancement with pumped-storage units," in: *2014 22nd Iranian Conference on Electrical Engineering (ICEE)*, 2014, pp. 820–825, doi:10.1109/IranianCEE.2014.6999649.

[67] M. Eidiani, "A new method of static available transfer capability computation in a high-penetration wind farm," *Majlesi Journal of Mechatronic Systems*, 2021, 10(2), pp. 29–38.

[68] Marc Champion, "The Future of Power Is Transcontinental, Submarine Supergrids," Bloomberg Businessweek, https://www.bloomberg.com/news/features/2021-06-09/future-of-world–energy-lies-in-uhvdc-transmission-lines, June 9, 2021.

[69] S. Lumbreras, H. Abdi, and R. Ramos, *Transmission Expansion Planning: The Network Challenges of the Energy Transition*, 2021. Springer Nature Switzerland AG, https://doi.org/10.1007/978-3-030-49428-5.

[70] G. D. A. Tinajero, M. Nasir, J. C. Vasquez, and J. M. Guerrero, "Comprehensive power flow modelling of hierarchically controlled AC/DC hybrid islanded microgrids," *International Journal of Electrical Power and Energy Systems*, 2021, 127, p. 106629, doi:10.1016/j.ijepes.2020.106629.

[71] H. Zhang and S. Wang, "A distributed dynamic power flow algorithm for an interconnected system containing two-terminal LCC-HVDC tie-line," *IEEE Access*, 2021, 9, pp. 28673–28683, doi:10.1109/ACCESS.2021.3058412.

[72] M. Eidiani and M. Kargar, "Frequency and voltage stability of the microgrid with the penetration of renewable sources," in: *2022 9th Iranian Conference on Renewable Energy & Distributed Generation (ICREDG)*, 2022, pp. 1–6, doi: 10.1109/ICREDG54199.2022.9804542.

[73] G. Ghardashi, M. Gandomkar, S. Majidi, M. Eidiani, and S. Dadfar, "Accuracy and speed improvement of microgrid islanding detection based on PV using frequency-reactive power feedback method," in: *2022 International Conference on Protection and Automation of Power Systems (IPAPS)*, 2022, pp. 1–8, doi:10.1109/IPAPS55380.2022.9763190.

[74] M. Eidiani, H. Zeynal, Z. Zakaria, and M. Shaaban, "A comprehensive study on the renewable energy integration using DIgSILENT," in: *2023 IEEE 3rd International Conference in Power Engineering Applications (ICPEA)*, 2023, 1570860205, 6–7.

[75] H. Zeynal, M. Eidiani, and D. Yazdanpanah, "Intelligent Substation Automation Systems for robust operation of smart grids," in: *2014 IEEE Innovative Smart Grid Technologies – Asia (ISGT ASIA)*, 2014, pp. 786–790, doi:10.1109/ISGT-Asia.2014.6873893.

[76] S. Gorman, "Electricity grid in US penetrated by spies," *The Wall Street Journal*, 2009, 8.

[77] INL/EXT-10-18381. "Vulnerability Analysis of Energy Delivery Control Systems," Idaho National Laboratory Idaho Falls, Idaho 83415, Sep. 2011.

[78] L. Xie, Y. Mo, and B. Sinopoli, "Integrity data attacks in power market operations," *IEEE Transactions on Smart Grid*, 2011, 2(4), pp. 659–666.

[79] G. Liang, S. R. Weller, J. Zhao, F. Luo, and Z. Y. Dong, "The 2015 Ukraine blackout: Implications for false data injection attacks," *IEEE Transactions on Power Systems*, 2017, 32, pp. 3317–3318.

[80] D. Case, "Analysis of the cyberattack on the Ukrainian power grid," Mar. 2016. https://media.kasperskycontenthub.com/wp-content/uploads/sites/43/2016/05/20081514/E-ISAC_SANS_Ukraine_DUC_5.pdf.

[81] A. Teixeira, S. Amin, H. Sandberg, K. H. Johansson, and S. S. Sastry, "Cybersecurity analysis of state estimators in electric power systems," in: *49th IEEE Conference on Decision and Control*, 2010, pp. 5991–5998.

[82] A. Ashok, M. Govindarasu, and V. Ajjarapu, "Online detection of stealthy false data injection attacks in power system state estimation," *IEEE Transactions on Smart Grid*, 2018, 9, pp. 1636–1646.

[83] S. Li, Y. Yılmaz, and X. Wang, "Quickest detection of false data injection attack in wide-area smart grids," *IEEE Transactions on Smart Grid*, 2015, 6(6), pp. 2725–2735.

[84] M.-Q. Tran, M. Elsisi, M.-K. Liu, *et al*, "Reliable deep learning and IoT-based monitoring system for secure computer numerical control machines against cyberattacks with experimental verification," *IEEE Access*, 2022, 10, pp. 23186–23197, doi:10.1109/ACCESS.2022.3153471.

[85] K. Yan, X. Liu, Y. Lu, and F. Qin, "A cyber-physical power system risk assessment model against cyberattacks," *IEEE Systems Journal*, 2022, 17, pp. 2018–2028, doi:10.1109/JSYST.2022.3215591.

[86] M. Eidiani, H. Zeynal, Z. Zakaria, and M. Shaaban, "Analysis of optimization methods applied for renewable energy integration," in: *2023 IEEE 3rd International Conference in Power Engineering Applications (ICPEA)*, 2023, 1570861924, 6–7.

[87] M. Eidiani, "Applying optimization techniques to develop a renewable energy supply map", in: Fathi, M., Zio, E., Pardalos, P.M. (eds.) *Handbook of Smart Energy Systems*. Springer, Cham. https://doi.org/10.1007/978-3-030-72322-4_61-1.

[88] M. Eidiani, "Integration of renewable energy sources," in: Fathi, M., Zio, E., Pardalos, P.M. (eds.) *Handbook of Smart Energy Systems,* 2022. Springer, Cham. https://doi.org/10.1007/978-3-030-72322-4_41-1

[89] W. Sicheng and G. Shan, "Available transfer capability analysis method of AC–DC power system based on security region," *The Journal of Engineering IET*, 2019, 2019(16), pp. 2386–2390.

Chapter 8

Advanced control approach for providing system strength

Myada Shadoul[1], Hassan Yousef[1], Rashid Al Abri[1,2] and Amer Al-Hinai[1]

Abstract

In this chapter, an adaptive controller for a fifth-order grid-connected system with several other control objectives will be presented and implemented. The purpose of the proposed adaptive fuzzy control (AFC) is to keep track of the target d−q current of the inverter where the uncertain dynamics of the PV grid-connected inverter system are estimated using a fuzzy logic approximator that uses the tracking error of the system. For the input–output feedback linearization control to be valid, the stability of the zero dynamics of the system is evaluated and proved to be stable.

8.1 Introduction

Today's world, where there is a rising need for energy, makes the strength of power systems more crucial than ever. The robustness of the energy grid guarantees a consistent and dependable supply of power. Power system strength will continue to be crucial in the effective and sustainable distribution of electricity as the globe advances toward renewable energy sources, assuring a dependable and secure energy future. The diversity and accessibility of the power sources, the infrastructure for transmission and distribution, and the efficiency of the control and protection systems are all elements that affect how strong a power system is. The strength and resilience of the energy grid depend on a well-designed and maintained power infrastructure. Effective planning, development, and operation are necessary, along with the establishment of advanced control and protection systems [1,2].

Renewable energy sources (RES) including solar photovoltaic (PV), wind turbines, as well as other sources, are incorporated into traditional power systems to avoid the high cost of creating new or extended power system infrastructure [3].

[1]Department of Electrical and Computer Engineering, Sultan Qaboos University, Oman
[2]Sustainable Energy Research Center, Sultan Qaboos University, Oman

Hence, power systems strength is heavily influenced by the integration of renewable energy sources (RES) [2]. DC−AC inverters are the final step in the PV system integration process. In electric power distribution systems, three-phase inverters are the most commonly used in grid-connected solar photovoltaic systems. Due to the electrical distinguishing features of the PV modules and the pulse width modulation technique (PWM) associated with the inverter, a PV system that is connected to the grid is a nonlinear system with unpredictable characteristics. Special care must be given to inverter designs and controls to improve the strength and reliability of power systems by preserving network reliability and maintaining the suitable performance of system parameters [4].

Voltage–source inverters (VSIs) and current–source inverters (CSIs) are the two most popular types of inverters in use [5]. In system modeling, the VSI and CSI introduce nonlinear dynamics and uncertainty. Such inverters can, however, provide controlled power factors and harmonic removal with a suitable design of control techniques [6,7]. Controlling such nonlinear systems is a difficult task; nevertheless, an approximate-based controller based on the fuzzy logic control concept provides a realistic solution.

For PV grid-connected inverters, many researchers proposed different types of controllers, including linear, nonlinear, adaptive, and model predictive controllers [8–14]. The fundamental disadvantage of traditional controller development methodologies is that they are dependent on the accessibility of system mathematical models as well as system variables. Intelligent control approaches such as neural networks, repetitive, autonomous, and fuzzy logic (FLCs) have been presented for nonlinear system control [15–21]. Intelligent controllers have the advantage of not relying on the system mathematical model and being able to handle a wide range of nonlinear and uncertain systems.

Different FLCs for PV grid-connected inverters were presented in [22–24]. The analysis of an FLC in real time to control the current and voltage of a grid-connected inverter was described in [23], where the findings demonstrated the FLC's capability to maintain the power factor. In [24], a grid inverter control employing FLC was proposed.

Furthermore, strategies for adaptive fuzzy control (AFC) based on fuzzy logic systems estimation features have been presented and implemented in a selection of nonlinear systems, such as power systems and drive applications [25–31].

The grid-connected PV inverter system is nonlinear with imprecise variables. Due to the unavailability of results in adaptive fuzzy approximation-based control, the authors suggest an AFC technique for PV grid-connected inverter systems. Where, the use of FLCs and AFC in controlling PV grid-connected inverters can contribute to the strength and reliability of power systems. The suggested controller utilizes the fuzzy logic system estimation ability and the principle of feedback linearization.

In this chapter, an adaptive controller for a fifth-order grid-connected system with several other control objectives will be presented and implemented. The purpose of the proposed AFC is to keep track of the target $d-q$ current of the inverter where the uncertain dynamics of the PV grid-connected inverter system are estimated using a fuzzy logic approximator that uses the tracking error of the system. For the input–output feedback linearization control to be valid, the stability of the zero dynamics of the system is evaluated and proved to be stable.

8.2 Fuzzy approximation controller for MIMO system

8.2.1 Input–output feedback linearization

We employ the concept of relative degree to develop a feedback linearization control, in which each output is repeatedly differentiated until at least one input occurs. The controlled system can be fully linearized if the system's order equals the system's relative degree. However, if the system's relative degree is smaller than its order, the controlled system is considered to be partially linearizable [32].

Define r_i to be the smallest integer such that at least one of the inputs appears in $y_j^{r_i}$. This means that

$$y_j^{r_i} = L_f^{r_i} h_j + L_{g_1} L_f^{r_1 - 1} h_1 u_1 + L_{g_2} L_f^{r_1 - 1} h_1 u_2 \tag{8.1}$$

with at least either $L_{g_1} L_f^{r_1 - 1} h_1 \neq 0$ or $L_{g_2} L_f^{r_1 - 1} h_1 \neq 0$. In (8.1), $L_f h(x) = \frac{\partial h}{\partial x} f(x)$ and $L_g h(x) = \frac{\partial h}{\partial x} g(x)$ stand for the Lie derivatives of $h(x)$ with respect to $f(x)$ and $g(x)$, respectively. The positive number r_i is termed as the relative degree of the corresponding output $h_i(x)$.

The input–output feedback linearizing control for a general nonlinear square MIMO system can be expressed as follows:

$$y^{(r)} = \alpha(x) + \beta(x)u \tag{8.2}$$

where $y^{(r)} = \begin{bmatrix} y_1^{(r_1)} \\ \vdots \\ y_m^{(r_p)} \end{bmatrix}, u = \begin{bmatrix} u_1 \\ \vdots \\ u_m \end{bmatrix}, r_1, r_2 \cdots r_m$ are the relative degrees of the outputs, m denotes the number of both inputs and outputs, and $\alpha(x) \in R^{m \times 1}$ and $\beta(x) \in R^{m \times m}$ are in general nonlinear functions.

In (8.2), the system functions $\alpha(x)$ and $\beta(x)$ particular values must be identified so the zero-tracking error is accomplished. In practice, $\alpha(x)$ and $\beta(x)$ may have unknown or inaccurate parameters. To avoid this limitation, the fuzzy systems' universal estimation ability is employed to estimate $\alpha(x)$ and $\beta(x)$. The concept of fuzzy adaptive approximation-based controller for general MIMO is described next.

8.2.2 General MIMO system fuzzy approximation controller

To estimate the nonlinear functions $\alpha_i(x)$ and $\beta_{ij}(x)$ in (8.2), a fuzzy logic system containing a singleton fuzzifier, product inference rules, and a weighted average defuzzifier is used. The approximated functions $\widehat{a}_i(x)$ and $\widehat{\beta}_{ij}(x), i = 1, 2 \ldots m$ and $j = 1, 2 \cdots m$ are generated using a fuzzy basis function (FBF) expansion $\xi(x)$ as

$$\widehat{a}_i(x) = \theta_i{}^T \xi(x) \tag{8.3}$$

$$\widehat{\beta}_{ij}(x) = \theta_{ij}{}^T \xi(x) \tag{8.4}$$

where $\theta_i \in R^{M \times 1}$, $\theta_{ij} \in R^{M \times 1}$ are the vectors of changeable parameters and $\xi(x) \in R^{M \times 1}$ represents the trajectory of fuzzy basis functions. The weighted-average defuzzifier is employed to construct the fuzzy basis function as

$$\xi_i(x) = \frac{\prod_{i=1}^{n} x_i \mu_{il}(x_i)}{\sum_{l=1}^{M} \left(\prod_{i=1}^{n} \mu_{il}(x_i) \right)} \tag{8.5}$$

where n stands for the number of states, M stands for the number of the If-Then rules, and the ith state in the lth rule membership function is $\mu_{il}(x_i)$.

When we replace $\alpha(x)$ and $\beta(x)$ in (8.2) with their conforming approximate (8.3) and (8.4), we obtained

$$y^{(r)} = \widehat{a}(x) + \widehat{\beta}(x)u \tag{8.6}$$

where $\widehat{a}(x) = \begin{bmatrix} \widehat{a}_1 \\ \vdots \\ \widehat{a}_m \end{bmatrix}$ and $\widehat{\beta}(x) = \begin{bmatrix} \widehat{\beta}_{11} & \cdots & \widehat{\beta}_{1m} \\ \vdots & \vdots & \vdots \\ \widehat{\beta}_{m1} & \cdots & \widehat{\beta}_{mm} \end{bmatrix}$.

Then the following control equation can be formed from (8.6) in terms of the fuzzy approximations $\widehat{a}(x)$ and $\widehat{\beta}(x)$

$$u = \widehat{\beta}^{-1}(x)(v - \widehat{a}(x)) \tag{8.7}$$

with

$$v = \begin{bmatrix} v_1 \\ \vdots \\ v_m \end{bmatrix} = y_{ref}^{(r)} + K\underline{e} \tag{8.8}$$

where $y_{ref}^r = \left[y_{ref1}^{(r_1)} \cdots y_{refm}^{(r_m)} \right]^T$, $K = \text{diag}\left[\underline{k}_1 \underline{k}_2 \cdots \underline{k}_m \right]$, $\underline{e} = \left[\underline{e}_1 \underline{e}_2 \cdots \underline{e}_m \right]^T$, $\underline{k}_i = \left[k_{0i} k_{1i} \cdots k_{(r_i-1)i} \right]$ and $\underline{e}_i = \left[e_i \dot{e}_i \cdots e_i^{(r_i-1)} \right]$ and $e_i = y_{ref1} - y_i$.

8.2.3   *Stability of the closed-loop*

By substituting (8.8) into (8.7) and the resulting equation is pre-multiplying by $\widehat{\beta}(x)$ to have

$$\widehat{\beta}(x)u = \left(y_{ref}^{(r)} - y^{(r)} \right) + K\underline{e} + y^{(r)} - \widehat{a}(x) \tag{8.9}$$

Using (8.2) in (8.9) to get the following error equation:

$$
\begin{bmatrix} e_1^{r_1} \\ \vdots \\ e_m^{r_m} \end{bmatrix} = -K\underline{e} + (\hat{a}(x) - a(x)) + \left(\hat{\beta}(x) - \beta(x)\right)u \tag{8.10}
$$

For the ith output the error equation can be expressed from (8.10) as

$$
e_i^{r_i} = -\underline{k}_i\underline{e}_i + \Delta a_i(x) + \sum_{j=1}^{m} \Delta\beta_{ij}(x)u_j \tag{8.11}
$$

where $\Delta a_i(x) = \hat{a}_i(x) - a_i(x)$ and $\Delta\beta_{ij}(x) = \hat{\beta}_{ij}(x) - \beta_{ij}(x)$ represent the fuzzy estimation errors. Equation (8.11) in state form can be written as

$$
\underline{\dot{e}} = A_i\underline{e}_i + \left[\Delta a_i(x) + \sum_{j=1}^{m} \Delta\beta_{ij}(x)u_j\right]b_i \tag{8.12}
$$

where $A_i = \begin{bmatrix} 0 & 1 & 0 & \cdots & 0 \\ 0 & 0 & 1 & \cdots & 0 \\ \vdots & \vdots & \vdots & \cdots & \vdots \\ 0 & 0 & 0 & 0 & 1 \\ -k_{(r_i-1)i} & k_{(r_i-2)i} & \cdots & \cdots & -k_{0i} \end{bmatrix}$ and $b_i = \begin{bmatrix} 0 \\ 0 \\ \cdots \\ 0 \\ 1 \end{bmatrix}$.

In terms of the optimal values of the adjustable parameters θ_i^* and θ_{ij}^*, the minimum fuzzy approximation error w_i can be defined as [25]

$$
w_i = \left[\hat{a}_i\left(x|\theta_i^*\right) - a_i(x)\right] + \sum_{j=1}^{m} \left[\hat{\beta}_{ij}\left(x|\theta_{ij}^*\right) - \beta_{ij}(x)\right]u_j \tag{8.13}
$$

The tracking error dynamics in (8.12) with parameter errors as $\varphi_{a_i} = \left(\theta_i - \theta_i^*\right)$ and $\varphi_{\beta_{ij}} = \left(\theta_{ij} - \theta_{ij}^*\right)$ then becomes

$$
\underline{\dot{e}} = A_i\underline{e}_i + b_i\left[w_i + \varphi_{a_i}^T\xi(x) + \sum_{j=1}^{m} \varphi_{\beta_{ij}}^T\xi(x)u_j\right] \tag{8.14}
$$

The parameter errors in (8.14) derivatives are given by

$$
\dot{\varphi}_{a_i} = \dot{\theta}_i \tag{8.15}
$$

$$
\dot{\varphi}_{\beta_{ij}} = \dot{\theta}_{ij} \tag{8.16}
$$

We utilize the next Lyapunov function to investigate the closed-loop stability and to determine the updating laws of the changeable parameters θ_i, θ_{ij}

$$
V_i = \frac{1}{2}\underline{e}_i^T P_i\underline{e}_i + \frac{1}{2\gamma_i}\varphi_{a_i}^T\varphi_{a_i} + \sum_{j=1}^{m} \frac{1}{2\gamma_{ij}}\varphi_{\beta_{ij}}^T\varphi_{\beta_{ij}} \tag{8.17}
$$

where γ_i and γ_{ij} are selected design variables and P_i is the solution of the below Lyapunov equation

$$A_i{}^T P_i + P_i A_i = -Q_i \tag{8.18}$$

where Q_i is the positive definite matrix.

The time derivative of (8.17) along the trajectories (8.14)–(8.16) is found as

$$\dot{V}_i = -\frac{1}{2}\underline{e}_i^T Q_i \underline{e}_i + \frac{1}{\gamma_i}\varphi_{\alpha_i}^T\left(\dot{\theta}_i + \gamma_i \underline{e}_i^T P_i b_i \xi(x)\right)$$

$$+ \left(\frac{1}{\gamma_{ij}}\sum_{j=1}^{m}\varphi_{\beta_{ij}}^T\dot{\theta}_{ij} + \underline{e}_i^T P_i b_i \sum_{j=1}^{m}\varphi_{\beta_{ij}}^T\xi(x)u_j\right) + \underline{e}_i^T P_i b_i w_i \tag{8.19}$$

Choosing the parameters' updating laws as

$$\dot{\theta}_i = -\gamma_i \underline{e}_i^T P_i b_i \xi(x) \tag{8.20}$$

$$\dot{\theta}_{ij} = -\gamma_{ij}\underline{e}_i^T P_i b_i \xi(x)u_j \tag{8.21}$$

then (8.19) becomes

$$\dot{V}_i = -\frac{1}{2}\underline{e}_i^T Q_i \underline{e}_i + \underline{e}_i^T P_i b_i w_i \tag{8.22}$$

From (8.17) and (8.22), the tracking and parameter errors are asymptotically stable, according to our findings.

8.3   PV grid-connected inverter adaptive fuzzy controller

8.3.1   *PV grid-connected inverter system model*

In Figure 8.1, a PV grid-connected inverter system is shown [5], the shown inverter is a single-stage inverter. A PV array is connected to a three-phase VSI through a DC link capacitor in this configuration. The VSI facilitates maximum power-point tracking (MPPT) in single-stage PV systems by transferring power to the grid and regulating the DC-link voltage. The grid is tied to the VSI AC side across an LC low pass-filter [33].

In Figure 8.1, the inverter is modeled using Park's transformation as (8.23)–(8.27), where the inverter voltage and current components are v_d, v_q, i_d, i_q the capacitor voltage components are v_{cd}, v_{cq}, the grid voltage and current components are $v_{gd}, v_{gq}, i_{gd}, i_{gq}$, the filter inductance and capacitance are L and C, and finally the grid angular frequency is ω:

$$\dot{i}_d = \omega i_q + \frac{1}{L}v_d - \frac{1}{L}v_{cd} \tag{8.23}$$

$$\dot{i}_q = -\omega i_d + \frac{1}{L}v_q - \frac{1}{L}v_{cq} \tag{8.24}$$

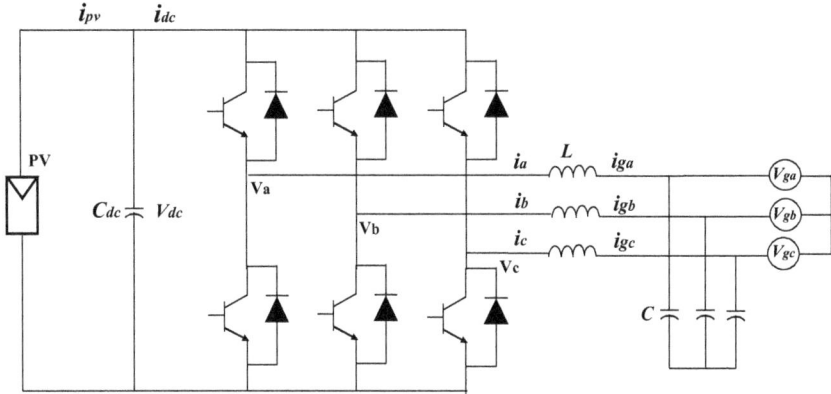

Figure 8.1   PV grid-connected inverter with LC filter

$$\dot{v}_{cd} = \omega v_{cq} + \frac{1}{C}i_d - \frac{1}{C}i_{gd} \tag{8.25}$$

$$\dot{v}_{cq} = -\omega v_{cd} + \frac{1}{C}i_q - \frac{1}{C}i_{gq} \tag{8.26}$$

$$\dot{v}_{dc} = \frac{i_{pv}}{C_{dc}} - \frac{3}{2C_{dc}V_{dc}}\left(v_{gd}i_{gd} + v_{gq}i_{gq}\right) \tag{8.27}$$

Hence, the system state-space model can be formed using (8.23)–(8.27) as (8.28)

$$\dot{x} = \begin{bmatrix} \omega x_2 - \dfrac{1}{L}x_3 \\ -\omega x_1 - \dfrac{1}{L}x_4 \\ \omega x_4 + \dfrac{1}{C}x_1 \\ -\omega x_3 + \dfrac{1}{C}x_2 \\ -\dfrac{3}{2C_{dc}x_5}\left(v_{gd}i_{gd} + v_{gq}i_{gq}\right) \end{bmatrix} + \begin{bmatrix} \dfrac{1}{L} \\ 0 \\ 0 \\ 0 \\ 0 \end{bmatrix}u_1 + \begin{bmatrix} 0 \\ \dfrac{1}{L} \\ 0 \\ 0 \\ 0 \end{bmatrix}u_2 + \begin{bmatrix} 0 \\ 0 \\ -\dfrac{1}{C}i_{gd} \\ -\dfrac{1}{C}i_{gq} \\ \dfrac{i_{pv}}{C_{dc}} \end{bmatrix} \tag{8.28}$$

where $x \in R^{n=5}$ is the state vector defined by (8.29) and the control inputs u_1 and u_2 are described by (8.30)

$$x = \begin{bmatrix} x_1 \\ x_2 \\ x_3 \\ x_4 \\ x_5 \end{bmatrix} = \begin{bmatrix} i_d \\ i_q \\ v_{cd} \\ v_{cq} \\ v_{dc} \end{bmatrix} \tag{8.29}$$

$$u = \begin{bmatrix} u_1 \\ u_2 \end{bmatrix} = \begin{bmatrix} v_d \\ v_q \end{bmatrix} \tag{8.30}$$

The purpose of the control is to design an adaptive fuzzy controller so the inverter currents i_d and i_q expressed by (8.31) will keep following specified target currents expressed by (8.32):

$$y = \begin{bmatrix} y_1 \\ y_2 \end{bmatrix} = \begin{bmatrix} i_d \\ i_q \end{bmatrix} \tag{8.31}$$

$$y_{ref} = \begin{bmatrix} i_{dref} \\ i_{qref} \end{bmatrix} = \begin{bmatrix} y_{ref1} \\ y_{ref2} \end{bmatrix} \tag{8.32}$$

It is worth mentioning that the inverter targets currents i_{dref} and i_{qref} are precisely related to the active and reactive power requirement of the inverter P_{ref} and Q_{ref} through the relations (8.33)

$$\begin{bmatrix} i_{dref} \\ i_{qref} \end{bmatrix} = \frac{2}{3} \begin{bmatrix} v_{gd} & v_{gq} \\ -v_{gq} & v_{gd} \end{bmatrix}^{-1} \begin{bmatrix} P_{ref} \\ Q_{ref} \end{bmatrix} \tag{8.33}$$

The state model of the system (8.28) and (8.31) could be written as

$$\dot{x} = f(x) + g_1(x)u_1 + g_2(x)u_2 + d, y = h(x) \tag{8.34}$$

where $f(x)$, $g_1(x)$, $g_2(x)$, and d are as defined in (8.28) and $h(x) = \begin{bmatrix} y_1 \\ y_2 \end{bmatrix} = \begin{bmatrix} h_1(x) \\ h_2(x) \end{bmatrix}$ is as defined in (8.31).

The input–output feedback linearization technique presented in Section 8.2 can be used to turn the model of the PV grid-connected inverter (8.34) into a feedback linearizable form. The nonlinear system dynamics (8.34) are converted into decoupled linear subsystems using a nonlinear control signal in this approach [32]. Next, the feedback linearization for the PV grid-connected inverter is presented.

8.3.2   Input–output feedback linearization for PV grid-connected inverter system

For the PV grid-connected inverter system, it can be shown that the relative degree $r_1 = r_2 = 1$ and (8.34) can be expressed it in the below format

$$\begin{bmatrix} \dot{y}_1 \\ \dot{y}_2 \end{bmatrix} = \alpha(x) + \beta(x) \begin{bmatrix} u_1 \\ u_2 \end{bmatrix} \tag{8.35}$$

where

$$\alpha(x) = \begin{bmatrix} L_f h_1(x) \\ L_f h_2(x) \end{bmatrix} = \begin{bmatrix} \omega x_2 - \dfrac{1}{L} x_3 \\ -\omega x_1 - \dfrac{1}{L} x_4 \end{bmatrix} \tag{8.36}$$

and

$$\beta(x) = \begin{bmatrix} L_{g_1}h_1 & L_{g_2}h_1 \\ L_{g_1}h_2 & L_{g_2}h_2 \end{bmatrix} = \begin{bmatrix} \dfrac{1}{L} & 0 \\ 0 & \dfrac{1}{L} \end{bmatrix} \tag{8.37}$$

The feedback linearizing control (8.38) gives the linear closed-loop system (8.39), and the external signals v_1 and v_2 are selected to guarantee asymptotic of the reference y_{ref} by the output vector y

$$\begin{bmatrix} u_1 \\ u_2 \end{bmatrix} = \beta^{-1}(x) \begin{bmatrix} v_1 - \alpha_1 \\ v_2 - \alpha_2 \end{bmatrix} \tag{8.38}$$

$$\begin{bmatrix} \dot{y}_1 \\ \dot{y}_2 \end{bmatrix} = \begin{bmatrix} v_1 \\ v_2 \end{bmatrix} \tag{8.39}$$

With describing the tracking error as

$$e = \begin{bmatrix} e_1 \\ e_2 \end{bmatrix} = \begin{bmatrix} y_{ref1} - y_1 \\ y_{ref2} - y_2 \end{bmatrix} \tag{8.40}$$

the signals v_1 and v_2 are selected as (8.41) and then the dynamics of the tracking error is found as (8.42)

$$\begin{bmatrix} v_1 \\ v_2 \end{bmatrix} = \begin{bmatrix} k_{01} & 0 \\ 0 & k_{02} \end{bmatrix} \begin{bmatrix} e_1 \\ e_2 \end{bmatrix} + \begin{bmatrix} \dot{y}_{ref1} \\ \dot{y}_{ref2} \end{bmatrix} \tag{8.41}$$

$$\left. \begin{array}{l} \dot{e}_1 + k_{01}e_1 = 0 \\ \dot{e}_2 + k_{02}e_2 = 0 \end{array} \right\} \tag{8.42}$$

where the coefficients k_{01} and k_{02} are design parameters chosen in order for the characteristic polynomials of (8.42) are Hurwitz and consequently ensuring that the tracking errors e_1 and e_2 asymptotically converging to zero [34].

For the introduced PV grid-connected inverter system, because $r_1 + r_2 = 2 = m < n$, hence the inverter current control is partially linearizable. The feedback linearization control (8.38) and (8.39) is valid if the system zero dynamics, that do not change as a result of feedback linearization, are stable [35]. The zero dynamics of the PV grid-connected inverter system can be found as follows. Define a transformed state $z^T = [\bar{z}_1 \quad \bar{z}_2]^T$ which represents a diffeomorphism of the state variables x where $\bar{z}_1^T = [h_1(x)h_2(x)] = [z_{11}(x)z_{12}(x)]$ and $\bar{z}_2^T = [z_{21}(x) \, z_{22}(x) \, z_{23}(x)]$. In these coordinates the system (8.34) is transformed to:

$$\left. \begin{array}{l} \dot{z}_{11}(x) = a_1(z) + b_1(z)u \\ \dot{z}_{12}(x) = a_2(z) + b_2(z)u \end{array} \right\} \tag{8.43}$$

where $a_i(z) = L_f h_i(x)$ and $b_i(z) = [L_{g_1}h_i \; L_{g_2}h_i], i = 1, 2$. Note that (8.43) is identical to (8.35). The internal dynamics of the transformed state $\bar{z}_2^T$ can be written as

$$\dot{z}_{2j}(x) = L_f z_{2j}(x) + L_{g_1} z_{2j}(x) u_1 + L_{g_2} z_{2j}(x) u_2 + L_d z_{2j}(x) \tag{8.44}$$

The condition to find $\bar{z}_2$ is that (8.44) to be independent of the linearizing control u_1 and u_2 [36], that is

$$\left. \begin{aligned} L_{g_1} z_{2j} &= \frac{\partial z_{2j}(x)}{\partial x} g_1 = 0 \\ L_{g_2} z_{2j} &= \frac{\partial z_{2j}(x)}{\partial x} g_2 = 0 \end{aligned} \right\}, j = 1, ..3 \tag{8.45}$$

It can be shown that the conditions (8.45) are satisfied with the following choice

$$\bar{z}_2 = \begin{bmatrix} \frac{1}{2} C_{dc} x_5^2 \\ x_3 \\ x_4 \end{bmatrix} \tag{8.46}$$

Using (8.46) in (8.44) to find the PV grid-connected inverter system internal dynamics as

$$\dot{z}_{21}(x) = L_f z_{21}(x) + L_d z_{2j}(x) = \frac{\partial z_{21}(x)}{\partial x_5}(f_5 + d_5) = C_{dc} x_5(f_5 + d_5) \tag{8.47}$$

$$\dot{z}_{22}(x) = L_f z_{22}(x) = \frac{\partial z_{22}(x)}{\partial x_3}(f_3 + d_3) = (f_3 + d_3) \tag{8.48}$$

$$\dot{z}_{23}(x) = L_f z_{23}(x) = \frac{\partial z_{23}(x)}{\partial x_4}(f_4 + d_4) = (f_4 + d_4) \tag{8.49}$$

where

$$\left. \begin{aligned} (f_3 + d_3) &= \omega x_4 + \frac{1}{C} x_1 - \frac{1}{C} i_{gd} \\ (f_4 + d_4) &= -\omega x_3 + \frac{1}{C} x_2 - \frac{1}{C} i_{gq} \\ (f_5 + d_5) &= \frac{i_{pv}}{C_{dc}} - \frac{3}{2 C_{dc} x_5}(v_{gd} i_{gd} + v_{gq} i_{gq}) \end{aligned} \right\} \tag{8.50}$$

If $0 \in R^5$ represents a stability point of the un-driven system, then the zero dynamics are given by

$$\left. \begin{aligned} \dot{z}_{21}(0) &= -\frac{3}{2}(v_{gd} i_{gd} + v_{gq} i_{gq}) \\ \dot{z}_{22}(0) &= -\frac{1}{C} i_{gd} \\ \dot{z}_{23}(0) &= -\frac{1}{C} i_{gq} \end{aligned} \right\} \tag{8.51}$$

which represents a stable system. Hence, the zero dynamics are stable, and the feedback linearizable control given by (8.38) and (8.39) is valid.

For PV grid-connected inverter systems, the accurate system functions values $\alpha(x)$ and $\beta(x)$ in (8.36) and (8.37) should be known to achieve zero tracking error. In practice, the PV grid-connected inverter system variables may be unknown or inaccurate. To address this shortcoming, the presented adaptive fuzzy controller is used to estimate $\alpha(x)$ and $\beta(x)$ which represent the system nonlinear functions. Next the adaptive fuzzy controller for the PV grid-connected inverter is presented.

8.3.3 PV grid-connected inverter adaptive fuzzy controller

The suggested AFC established on partial feedback linearization given by (8.3), (8.4), (8.6), (8.20), and (8.21) is applied to the introduced PV grid-connected inverter system in Figure 8.1. Figure 8.2 depicts the suggested controller's block diagram. As illustrated in the block diagram, the grid voltages and currents are converted from an abc frame to a dq frame. With the calculation beginning from a selected set of preliminary values of θ_i and θ_{ij}, the unidentified controllers'

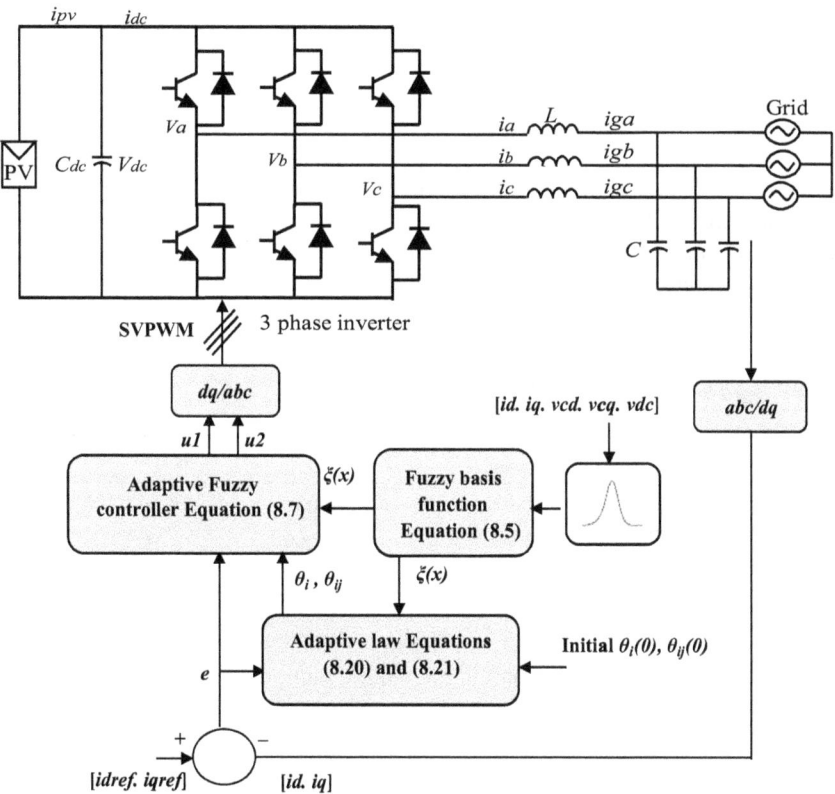

Figure 8.2   Adaptive fuzzy approximation controller block diagram

parameters θ_i and θ_{ij} of the controlled system were approximated utilizing the control laws in (8.20) and (8.21). The control signals were then generated using AFC low in (8.7). Then, by utilizing space vector PWM (SVPWM) the PWM signal will be obtained to run the inverter. To execute the suggested AFC, fuzzy sets F_k^i must be chosen in which $k = 1, 2, \cdots 5$ and $i = 1, 2, \ldots P$, P is the fuzzy sets numbers. To find the vector of fuzzy basis functions given in (8.5), fuzzy sets are used. The fuzzy basis functions for each state of the system are generated using three Gaussian fuzzy sets: small (S), zero (Z), and big (B). The membership functions of these fuzzy sets define them. The generic form of Gaussian-type membership functions is provided by

$$\mu_{F_k^i}(x_k) = \exp\left(-\frac{\left(x_k - \bar{x}_k^i\right)^2}{\sigma_k^i}\right) \tag{8.52}$$

The parameters of the membership functions are displayed in Table 8.1, where $x_i(0)$, is selected as the center of the related MF, while the remaining parameters in the constraint sets are determined at arbitrary [25]. Figure 8.3 depicts the membership functions for the x_1 as an example.

For the PV grid-connected inverter system, the AFC law (8.6) becomes

$$\begin{bmatrix} u_1 \\ u_2 \end{bmatrix} = \begin{bmatrix} \hat{\beta}_{11}(x) & \hat{\beta}_{12}(x) \\ \hat{\beta}_{21}(x) & \hat{\beta}_{22}(x) \end{bmatrix}^{-1} \begin{bmatrix} -\hat{a}_1(x) + v_1 \\ -\hat{a}_2(x) + v_2 \end{bmatrix} \tag{8.53}$$

where $\hat{a}_i$ and $\hat{\beta}_{ij}$ are produced using the fuzzy estimate (8.3) and (8.4) together with the update laws (8.20) and (8.21). Equation (8.41) is used to choose $[v_1 \ v_2]^T$. The matrices A_i, and $b_i, i = 1, 2$ in (8.12) are transformed into $A_1 = -k_{01}$, $b_1 = 1$, $A_2 = -k_{02}$ and $b_2 = 1$. The control gains k_{01} and k_{02} must be high sufficient to

Table 8.1　The membership functions' variables

State↓Fuzzy set→	S	Z	B
x_1	$\bar{x}_1^S = 1{,}100$	$\bar{x}_1^Z = 1{,}400$	$\bar{x}_1^B = 1{,}700$
	$\sigma_1^S = 20{,}000$	$\sigma_1^Z = 20{,}000$	$\sigma_1^B = 20{,}000$
x_2	$\bar{x}_2^S = -10$	$\bar{x}_2^Z = 0$	$\bar{x}_2^B = 10$
	$\sigma_2^S = 6$	$\sigma_2^Z = 6$	$\sigma_2^B = 6$
x_3	$\bar{x}_3^S = 320$	$\bar{x}_3^Z = 340$	$\bar{x}_3^B = 360$
	$\sigma_3^S = 100$	$\sigma_3^Z = 100$	$\sigma_3^B = 100$
x_4	$\bar{x}_4^S = -10$	$\bar{x}_4^Z = 0$	$\bar{x}_4^B = 10$
	$\sigma_4^S = 6$	$\sigma_4^Z = 6$	$\sigma_4^B = 6$
x_5	$\bar{x}_5^S = 1{,}200$	$\bar{x}_5^Z = 1{,}290$	$\bar{x}_5^B = 1{,}380$
	$\sigma_5^S = 2{,}000$	$\sigma_5^Z = 2{,}000$	$\sigma_5^B = 2{,}000$

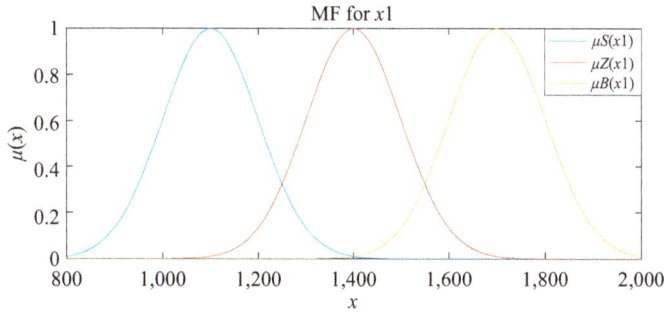

Figure 8.3 x_1 Membership function

achieve starting stability due to the iterative nature of the adaptation process. A trial-and-error approach and sensible limitations are used to determine the design parameters γ_i and γ_{ij} [25].

8.4   Simulation situations and results

A three-phase grid-connected PV inverter system with the specifications indicated in Table 8.2 [33] was used to evaluate the effectiveness of the suggested controller. For the various operating situations below, the suggested controller is simulated using MATLAB®/SIMULINK®.

8.4.1   Situation I: unity power factor

In this situation, simulating the regulation of the power factor to unity is done by choosing the target current components of the grid as follows: $i_{qref} = 0$ A and $i_{dref} = 1,400$ A. Figure 8.4 depicts the current and voltage of the grid, illustrating that they are in phase, indicating that the PF is unity. Figure 8.5(a, b) depicts the tracking of current i_q and to its target value i_{qref} in about 5 ms and tracking of the current i_d to its target respectively. In Figure 8.6(a, b), the control signals u_1 and u_2 are presented, and it can be seen that they are both bounded. Hence, the achieved results illustrate that the suggested AFC offers excellent current tracking performance.

8.4.2   Situation II: tracking of reactive current changes

The system's reaction to reactive current changes is shown in this situation. At $t = 0.05$ s, the target of the grid current components i_{qref} is considered to have a difference of 200 A, which equates to a power factor shift to 0.98. Figure 8.7 displays (~ 1 ms)   phase change between the current and voltage waveforms, indicating that the power factor is changed with the change in i_{qref}. Furthermore, Figure 8.8(a) depicts the grid current i_q tracking its target i_{qref}. The grid current i_d tracking to its target is shown in Figure 8.8(b) and is unaffected by the i_{qref} step

Table 8.2   Variables of the controlled system

Variable	Value
The voltage of the PV array V_{dc}	1,290 V
Capacitor of the DC link	5,000 µF
Inductance of the filter L	100 µH
Capacitance of the filter C	369 µF
Voltage of the grid	415 V
Frequency of the grid	50 Hz
Parameters k_{ij}	$k_{01} = 10,000$, $k_{02} = 10,000$
Design parameters γ_i and γ_{ij}	$\gamma_1 = 4$, $\gamma_2 = 0.01$
	$\gamma_{11} = 0.01$, $\gamma_{12} = 0.1$
	$\gamma_{21} = 0.1$ and $\gamma_{22} = 1$

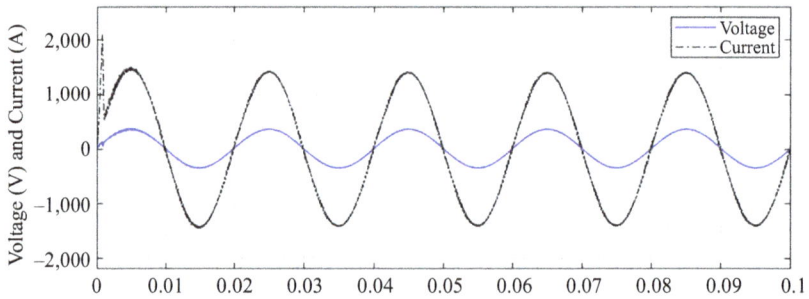

Figure 8.4   Unity power factor: waveforms of voltage and current

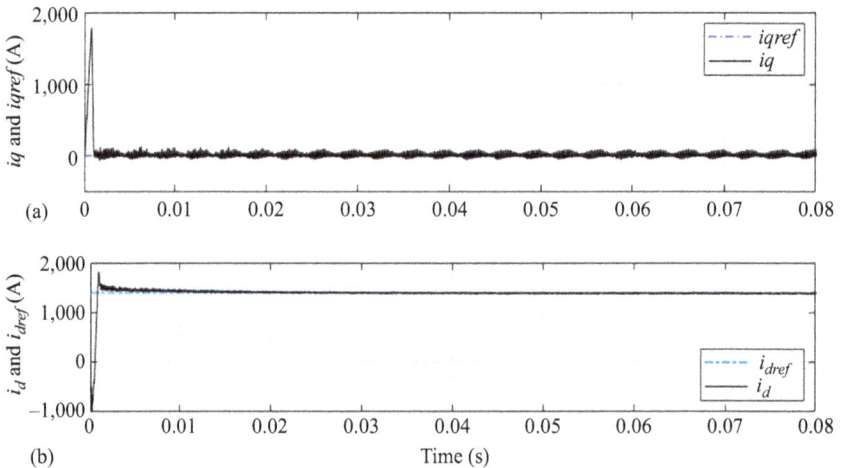

Figure 8.5   Unity power factor: (a) reactive current and its target. (b) Active current and its target.

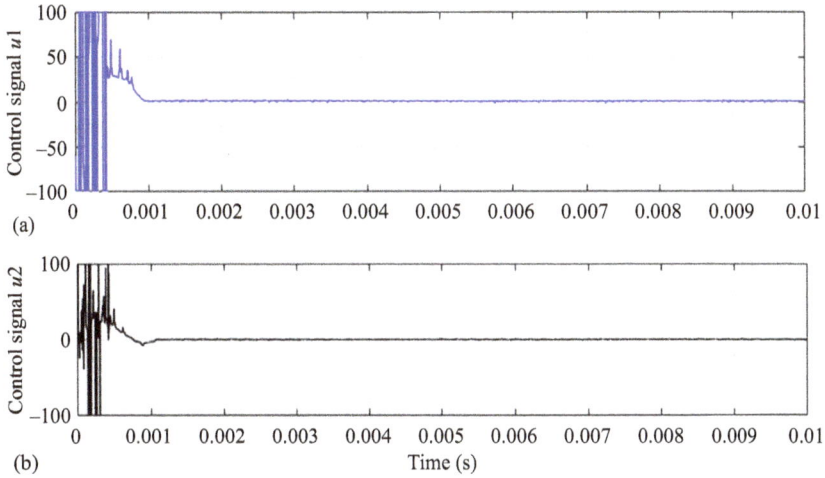

Figure 8.6   Unity power factor: (a) control signal u_1. (b) Control signal u_2.

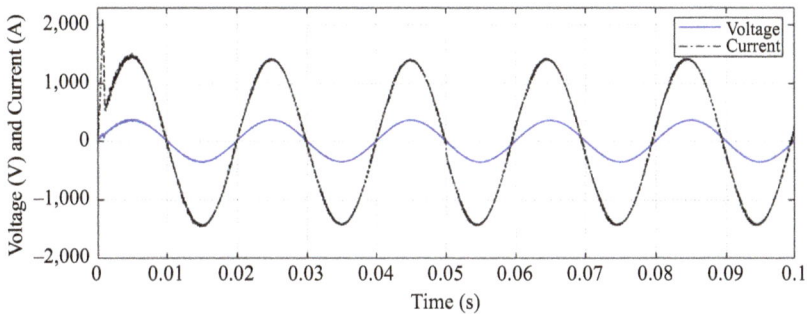

Figure 8.7   Tracking of reactive current changes: waveforms of voltage and current

change. In Figure 8.8, a high oscillation in current i_q can be noted, this oscillation is caused by current control strategy and loops coupling. Figure 8.9 depicts the power factor changes caused by a step change in i_{qref}.

8.4.3   Situation III: tracking of active current changes

To test the efficiency of the suggested controller in tracking the changes in active current, the target active current i_{dref} is considered to have a difference of 400 A (which occurs at $t = 0.045$ s). Voltage and current waveforms with active current target changes are shown in Figure 8.10, confirming that the new active current is being tracked. The current component i_d and i_{dref} are shown in Figure 8.11(b), verifying the tracking of the required change of the active current. Moreover, the

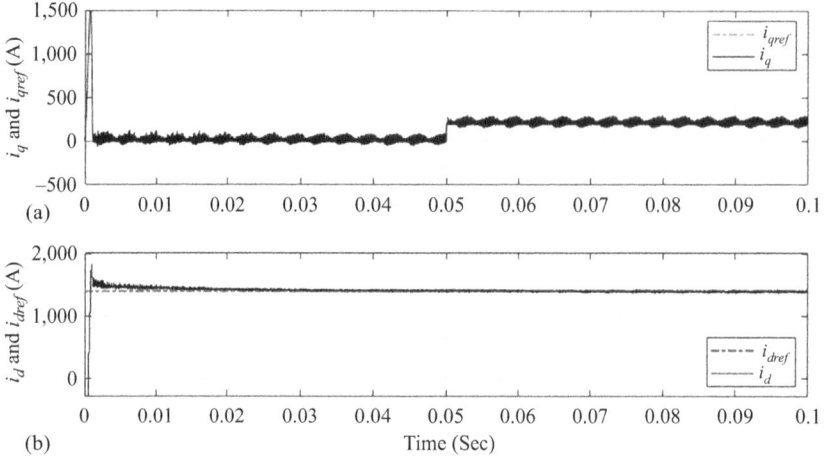

Figure 8.8   Tracking of reactive current changes: (a) reactive current and its target. (b) Active current and its target.

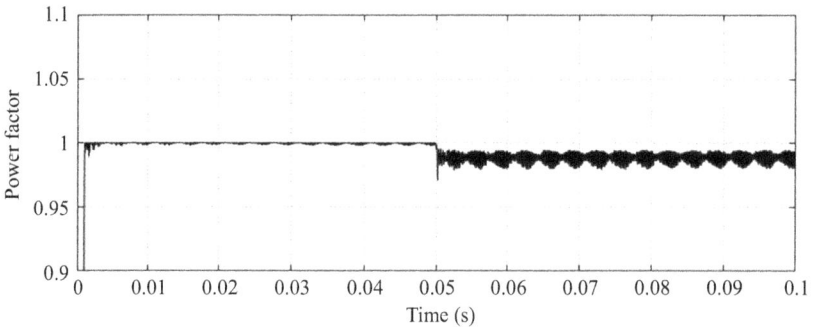

Figure 8.9   Tracking of reactive current changes: power factor

controller performance shows decoupling between the changes in the current component i_d and current components i_q.

8.4.4   Situation IV: robust tracking

The operation of the proposed controller is examined for different uncertainties in the PV grid-connected inverter system parameters, namely the filter parameters L and C. Simulation results for 20% variation in L and -20% variation in C are depicted in Figures 8.12 and 8.13 correspondingly, with $i_{dref} = 1,400$ A and $i_{qref} = 0$. The performance of the PV grid-connected inverter system under these uncertainties proves that the proposed adaptive fuzzy approximation controller is capable of achieving good tracking performance.

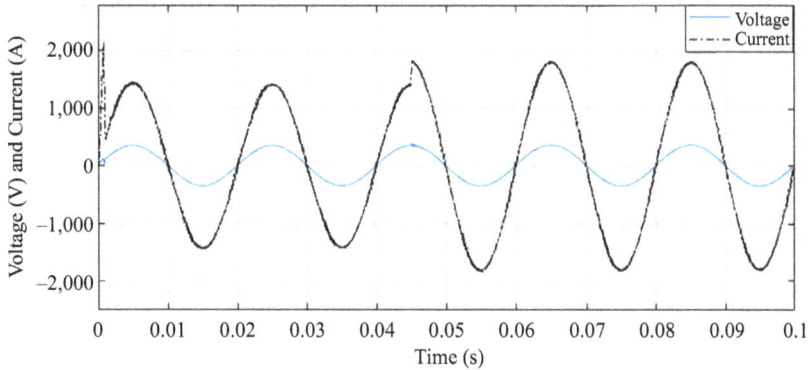

Figure 8.10   Tracking of active current changes: voltage and current waveforms

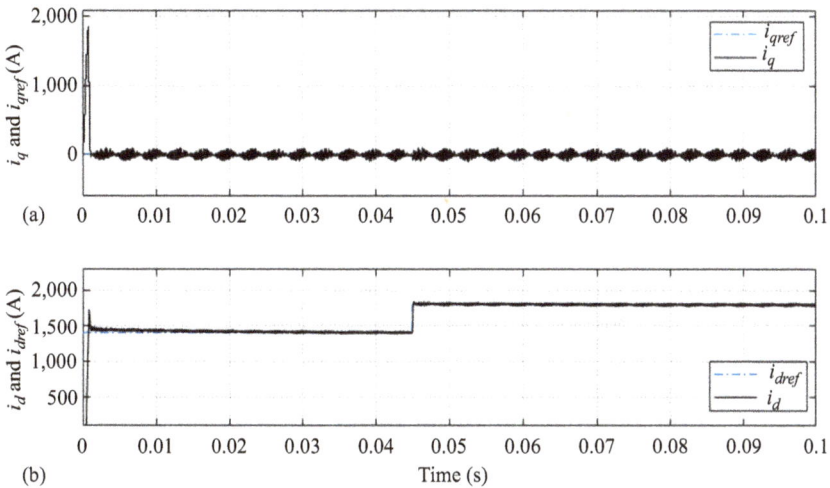

Figure 8.11   Tracking of active current changes: (a) reactive current and its target. (b) Active current and its target.

A comparison of the tracking functionality of the proposed AFC, a newly designed PI controller based on the PSO algorithm [37], and other controllers given in [8] and [38] was carried out. The chosen performance indicators are selected as the settling time and rise time. Table 8.3 shows these performance indicators for the unity power factor tracking case. Table 8.3 demonstrates that the suggested controller surpasses the PI based on the PSO algorithm controller and other controllers in respect of tracking the needed reference signal with a quicker settling time and shorter rising time.

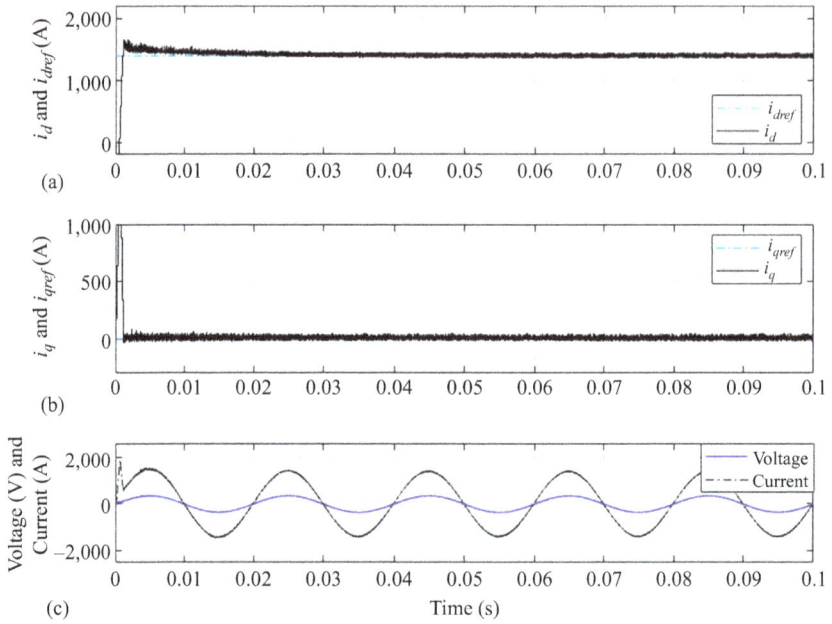

Figure 8.12   Simulation results for 20% variation in L: (a) active current and its target. (b) Reactive current and its target. (c) Waveforms of the voltage and current.

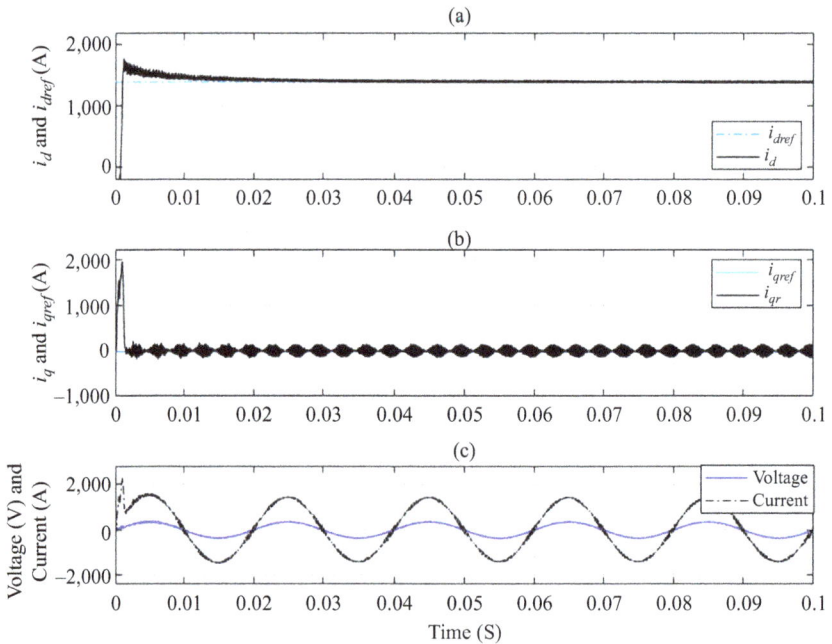

Figure 8.13   Simulation results for −20% variation in C: (a) active current its target. (b) Reactive current and its target. (c) Waveforms of the voltage and current.

Table 8.3 Performance measures

Controller	Settling time (s)	Rise time (s)
PI-PSO controller [37]	0.1853	0.112
Takagi–Sugeno–Kang probabilistic fuzzy neural network [8]	0.3	–
A finite control set model predictive current control (FCS-MPCC) [38]	0.08	–
Proposed AFC	0.01997	0.00029

8.5 Research gaps and future work

There are several research gaps in the design of adaptive fuzzy controllers for grid-connected inverters, which is an active research topic. Following are a few of the topic's research gaps, and potential future study directions:

(A) Research gaps:
  • There is a lack of research on how well adaptive fuzzy control works to lower grid voltage and frequency variations.
  • The effect of design factors on the stability and robustness of adaptive fuzzy controllers has received little attention.
  • There is a lack of research on combining adaptive fuzzy control with other a grid-connected inverter control methods.

(B) Future work:
  • Exploring the application of different types of fuzzy logic systems like type-2 fuzzy logic systems, interval type-2 fuzzy logic systems for adaptive control of grid-connected inverters.
  • Design and test adaptive fuzzy controller for grid-connected inverter with LCL filter.
  • More different operating scenarios for the proposed controllers can be validated.
  • Examining how various load types and profiles affect the effectiveness of adaptive fuzzy controllers.
  • The proposed AFC methods can be experimentally validated.
  • Creating adaptive fuzzy controllers that can efficiently reduce the effect of renewable energy sources on the inverter connected to the grid.

8.6 Conclusions

In this chapter, adaptive fuzzy-based control for a PV grid-connected inverter system with an LC filter is suggested. The suggested AFC is designed to help improve the strength and reliability of power systems. The fuzzy system's estimation ability and the feedback linearization concept are employed to design the controller. As a nonlinear MIMO system, the controlled inverter system is modeled. The controlled system analysis proved that the system is partially linearizable and its zero dynamics are stable. The adaptive fuzzy control law was established by utilizing a fuzzy system to approximate the unknown nonlinear functions that occur in state feedback linearizing control. Closed-loop stability and control parameter update laws are obtained using Lyapunov analysis. The suggested AFC was executed and evaluated using MATLAB/SIMULINK for the PV grid-connected inverter system for multiple operating situations such as power factor unity and change tracking, active current change tracking, and uncertainty in system parameters tracking. The proposed method's simulation results show that it has good tracking performance in a variety of operational cases. Furthermore, the proposed controller can deal with filter parameter uncertainty. The current components i_d and i_q decoupling are illustrated in all simulation results. In addition, a comparison of the suggested AFC with a PI controller based on the PSO algorithm demonstrates that the suggested AFC outperforms the PI controller in various tracking performance analyses, including settling time and rising time.

References

[1]   J. Liu, W. Yao, J. Wen, *et al.*, "Impact of power grid strength and PLL parameters on stability of grid-connected DFIG wind farm," *IEEE Transactions on Sustainable Energy*, vol. 11, pp. 545–557, 2019.

[2]   H. Urdal, R. Ierna, J. Zhu, C. Ivanov, A. Dahresobh, and D. Rostom, "System strength considerations in a converter dominated power system," *IET Renewable Power Generation*, vol. 9, pp. 10–17, 2015.

[3]   A. Zahedi, "A review of drivers, benefits, and challenges in integrating renewable energy sources into electricity grid," *Renewable and Sustainable Energy Reviews*, vol. 15, pp. 4775–4779, 2011.

[4]   T. Dragičević, X. Lu, J. C. Vasquez, and J. M. Guerrero, "DC microgrids— Part II: a review of power architectures, applications, and standardization issues," *IEEE Transactions on Power Electronics*, vol. 31, pp. 3528–3549, 2015.

[5]   K. Zeb, W. Uddin, M. A. Khan, *et al.*, "A comprehensive review on inverter topologies and control strategies for grid connected photovoltaic system," *Renewable and Sustainable Energy Reviews*, vol. 94, pp. 1120–1141, 2018.

[6]   E. P. Wiechmann, P. Aqueveque, R. Burgos, and J. Rodríguez, "On the efficiency of voltage source and current source inverters for high-power drives," *IEEE Transactions on Industrial Electronics*, vol. 55, pp. 1771–1782, 2008.

[7]   Y. W. Li, "Control and resonance damping of voltage–source and current–source converters with $ LC $ filters," *IEEE Transactions on Industrial Electronics*, vol. 56, pp. 1511–1521, 2008.

[8]   F.-J. Lin, K.-C. Lu, T.-H. Ke, B.-H. Yang, and Y.-R. Chang, "Reactive power control of three-phase grid-connected PV system during grid faults

using Takagi–Sugeno–Kang probabilistic fuzzy neural network control," *IEEE Transactions on Industrial Electronics*, vol. 62, pp. 5516–5528, 2015.

[9]   P. P. Dash and M. Kazerani, "Dynamic modeling and performance analysis of a grid-connected current-source inverter-based photovoltaic system," *IEEE Transactions on Sustainable Energy*, vol. 2, pp. 443–450, 2011.

[10] M. Mahmud, M. Hossain, H. Pota, and N. Roy, "Robust nonlinear controller design for three-phase grid-connected photovoltaic systems under structured uncertainties," *IEEE Transactions on Power Delivery*, vol. 29, pp. 1221–1230, 2014.

[11] X. Bao, F. Zhuo, Y. Tian, and P. Tan, "Simplified feedback linearization control of three-phase photovoltaic inverter with an LCL filter," *IEEE Transactions on Power Electronics*, vol. 28, pp. 2739–2752, 2012.

[12] H. M. Hasanien, "An adaptive control strategy for low voltage ride through capability enhancement of grid-connected photovoltaic power plants," *IEEE Transactions on Power Systems*, vol. 31, pp. 3230–3237, 2015.

[13] X. Li, H. Zhang, M. B. Shadmand, and R. S. Balog, "Model predictive control of a voltage–source inverter with seamless transition between islanded and grid-connected operations," *IEEE Transactions on Industrial Electronics*, vol. 64, pp. 7906–7918, 2017.

[14] P. Monica and M. Kowsalya, "Control strategies of parallel operated inverters in renewable energy application: a review," *Renewable and Sustainable Energy Reviews*, vol. 65, pp. 885–901, 2016.

[15] P. M. De Almeida, J. L. Duarte, P. F. Ribeiro, and P. G. Barbosa, "Repetitive controller for improving grid-connected photovoltaic systems," *IET Power Electronics*, vol. 7, pp. 1466–1474, 2014.

[16] E. Harirchian and T. Lahmer, "Improved rapid visual earthquake hazard safety evaluation of existing buildings using a type-2 fuzzy logic model," *Applied Sciences*, vol. 10, p. 2375, 2020.

[17] Q. A. Tarbosh, Ö. Aydoğdu, N. Farah, *et al.*, "Review and investigation of simplified rules fuzzy logic speed controller of high performance induction motor drives," *IEEE Access*, vol. 8, pp. 49377–49394, 2020.

[18] M. Pushpavalli and N. Jothi Swaroopan, "KY converter with fuzzy logic controller for hybrid renewable photovoltaic/wind power system," *Transactions on Emerging Telecommunications Technologies*, vol. 12, p. e3989, 2020.

[19] S. Mumtaz, S. Ahmad, L. Khan, S. Ali, T. Kamal, and S. Z. Hassan, "Adaptive feedback linearization based neurofuzzy maximum power point tracking for a photovoltaic system," *Energies*, vol. 11, p. 606, 2018.

[20] M. Hosseinzadeh and F. R. Salmasi, "Power management of an isolated hybrid AC/DC micro-grid with fuzzy control of battery banks," *IET Renewable Power Generation*, vol. 9, pp. 484–493, 2015.

[21] E. Harirchian and T. Lahmer, "Developing a hierarchical type-2 fuzzy logic model to improve rapid evaluation of earthquake hazard safety of existing buildings," *Structures*, 2020, pp. 1384–1399.

[22]   G. Chen and T. T. Pham, *Introduction to Fuzzy Sets, Fuzzy Logic, and Fuzzy Control Systems*. London: CRC Press, 2000.

[23]   M. Hannan, Z. A. Ghani, A. Mohamed, and M. N. Uddin, "Real-time testing of a fuzzy-logic-controller-based grid-connected photovoltaic inverter system," *IEEE Transactions on Industry Applications*, vol. 51, pp. 4775–4784, 2015.

[24]   S. Muyeen and A. Al-Durra, "Modeling and control strategies of fuzzy logic controlled inverter system for grid interconnected variable speed wind generator," *IEEE Systems Journal*, vol. 7, pp. 817–824, 2013.

[25]   L. X. Wang and H. Ying, "Adaptive fuzzy systems and control: design and stability analysis," *Journal of Intelligent and Fuzzy Systems—Applications in Engineering and Technology*, vol. 3, p. 187, 1995.

[26]   Y.-J. Liu, W. Wang, S.-C. Tong, and Y.-S. Liu, "Robust adaptive tracking control for nonlinear systems based on bounds of fuzzy approximation parameters," *IEEE Transactions on Systems, Man, and Cybernetics—Part A: Systems and Humans*, vol. 40, pp. 170–184, 2009.

[27]   Y. Li, S. Tong, Y. Liu, and T. Li, "Adaptive fuzzy robust output feedback control of nonlinear systems with unknown dead zones based on a small-gain approach," *IEEE Transactions on Fuzzy Systems*, vol. 22, pp. 164–176, 2013.

[28]   H. A. Yousef and M. A. Wahba, "Adaptive fuzzy mimo control of induction motors," *Expert Systems with Applications*, vol. 36, pp. 4171–4175, 2009.

[29]   H. M. Nguyen, *Advanced Control Strategies for Wind Energy Conversion Systems*. Pocatello, ID: Idaho State University, 2013.

[30]   C. Zhou, D.-C. Quach, N. Xiong, *et al.*, "An improved direct adaptive fuzzy controller of an uncertain PMSM for web-based e-service systems," *IEEE Transactions on Fuzzy Systems*, vol. 23, pp. 58–71, 2014.

[31]   X. Liu, D. Zhai, J. Dong, and Q. Zhang, "Adaptive fault-tolerant control with prescribed performance for switched nonlinear pure-feedback systems," *Journal of the Franklin Institute*, vol. 355, pp. 273–290, 2018.

[32]   H. K. Khalil, *Nonlinear Systems*, Upper Saddle River, NJ: Michigan State University, 2002.

[33]   A. Yazdani, A. R. Di Fazio, H. Ghoddami, *et al.*, "Modeling guidelines and a benchmark for power system simulation studies of three-phase single-stage photovoltaic systems," *IEEE Transactions on Power Delivery*, vol. 26, pp. 1247–1264, 2010.

[34]   D. Lalili, A. Mellit, N. Lourci, B. Medjahed, and E. Berkouk, "Input output feedback linearization control and variable step size MPPT algorithm of a grid-connected photovoltaic inverter," *Renewable Energy*, vol. 36, pp. 3282–3291, 2011.

[35]   S. Sastry and M. Bodson, *Adaptive Control: Stability, Convergence, and Robustness*. New Jersey, NJ: Prentice-Hall, 1989.

[36]   J. K. Hedrick and A. Girard, "Control of nonlinear dynamic systems," Berkeley, CA: University of California, 2015.

[37] M. Roslan, A. Q. Al-Shetwi, M. Hannan, P. Ker, and A. Zuhdi, "Particle swarm optimization algorithm-based PI inverter controller for a grid-connected PV system," *PLoS One*, vol. 15, p. e0243581, 2020.

[38] Y. Zhao, A. An, Y. Xu, Q. Wang, and M. Wang, "Model predictive control of grid-connected PV power generation system considering optimal MPPT control of PV modules," *Protection and Control of Modern Power Systems*, vol. 6, pp. 1–12, 2021.

Chapter 9

The impact of renewable energy on voltage stability and fault level

Luca Vignali[1] and Alberto Berizzi[2]

Abstract

The transition to complete decarbonization is bringing about a total revolution in the electric power system paradigm. This chapter is focused on issues related to voltage stability and fault level decadence, proposing an effective procedure for short-circuit analysis including the latest generation of power electronic converters but always maintaining a steady-state approach. The computation tool is tested and exploited to perform studies on high-renewable energy source penetration grids to discuss improvements and strategies on planning and operation for future power systems.

9.1 Introduction

Power systems are experiencing an important period of transition characterized by an ever-growing need for energy together with an increasing renewable energy sources (RESs) penetration. This has given rise to issues never experienced before: among them, the progressive reduction of fault level due to limited short-circuit current contribution of converter-based generation, that could result in, among others, increased risk of voltage instability, protection system misoperation and commutation failure of High Voltage Direct Current (HVDC) systems. Given the negative effect on system strength, it is becoming more and more relevant to analyze the impact of RES on short-circuit power, in an attempt to investigate possible solutions so as to guarantee a good quality service and improve power system operation and security.

Accordingly, cutting-edge computation and analysis tools are necessary to study modern power systems and set up strategies against the aggravation of system strength in the presence of high-RES penetration. In particular, short-circuit analysis needs to be reconsidered, since RES can no longer be neglected in computations, and in particular the non-linear behavior of converters under fault conditions has to be focused on. In fact, tools used nowadays are becoming less and less

[1]Centro Elettrotecnico Sperimentale Italiano (CESI) SpA
[2]Politecnico di Milano, Energy Department

suitable, because they neglect RES contribution, underestimating fault currents; moreover, they also neglect possible negative interactions among converters.

Traditional converters are electrically comparable to voltage-controlled current sources; hence, their fault current is non-linearly dependent on voltage conditions, which means that system strength is not only fundamental in guaranteeing their stable operation but also to prevent perturbations from propagating over long distances and cause cascading disconnections of many converter-based generators and/ or HVDC systems. Cascading events, which are usually neglected in short-circuit analysis, are one of the most relevant issues in weak grids and, as RESs are becoming a larger share of global power generation, it is important they remain connected during faults both to avoid unacceptable power loss and to sustain the voltage.

Along these lines, new procedures to model non-linear RES fault contribution have been developed in the last years, presenting some methods based on time-domain dynamic analysis, such as [1,2], often also with the aid of commercial software packages, as in the case of [3–5]. Nevertheless, setting up the dynamic model of a large network is complex, and the dynamic approach cannot be applied to massive short-circuit computations, which is a necessary duty for Transmission System Operators (TSOs). A practical steady-state Root Mean Square (RMS) computation method is still needed, but it should be updated as in [6,7] to include converter models; this is particularly interesting, since it allows for keeping valid the superposition principle also in the case of RES, avoiding dynamic simulations.

In accordance with the same guidelines, some calculation methods have been developed in this chapter; then, after being tested to prove its validity, the tool has been used to perform operational planning studies so as to discuss the future of electric systems, comparing different converter technologies and investigating the best strategies to improve system strength, stability, and reliability. Specifically, the analysis has been carried out on Sicily's power system, in Italy, and the outcomes have been compared with those of a well-known dynamic simulation program. The tool has been intended to include also cascading events, such as RES disconnection and crowbar protection of wind turbines (WTs).

Finally, the new algorithm has been designed to include grid forming (GFM) converters as well. GFMs are still under development, and there are not many related studies, almost all focused on dynamic simulations and models, such as [8–10]. The GFM converters modeling in steady-state short-circuit analysis, the computation of short-circuit currents including this new converter technology, and the study of their impact on system strength are further new issues dealt with in the present chapter.

9.2 Highlights

The layout of this chapter will be dealt with below through the main steps of the analysis:

- A new algorithm is designed and implemented in MATLAB® to perform steady-state RMS fault analysis including traditional grid following (GFL) and GFM converters.

- The computation tool is used to test different fault ride-through (FRT) strategies, investigating operational improvements and possible updates of the Italian Grid Code.
- The impact of RES on fault level and system strength is studied, considering Sicily's power system in Italy; moreover, a hypothetic 100% RES Sicilian grid is modeled to estimate future fault level reduction and an attempt will be made to improve it with RES.
- GFM impact on fault level is studied, examining system strength benefits and comparing with GFL converters.

9.3   Power system strength

The system strength concept and related issues are introduced in this section [11,12], being among the key targets of the analysis. Power system strength can be defined as a characteristic of electrical grids related to the voltage change following a fault or disturbance. In particular, system strength is the ability of power systems to maintain and control the voltage waveform at any given location, both in normal and in emergency conditions.

The system strength is a key factor to assess how well the power system can return to normal operation following a perturbation, or differently, how quickly the power system voltage waveform can be restored to pre-disturbance conditions. Power systems with good system strength can maintain stable voltages following changes in power flows and will be more tolerant to variations and perturbations, recovering better from major disturbances such as faults and sudden loss of equipment.

The system strength is defined at a given location and it is proportional to the fault level at that location; hence, it is clear that the power generation technology has a major impact on it. Synchronous machines are generally a positive contributor to system strength given their inherent high short-circuit current, which depends on their physics: a synchronizing torque is always available in a synchronous machine. On the contrary, converter-based RES generation has generally been considered as a negative contributor to system strength, mainly because GFL technologies were contemplated; their synchronization capability is not inherent, but it is established by the converter control. The emerging of new technologies such as GFM inverters or the upgrading of FRT control strategies is expected to be mitigating this effect, as will be discussed in this analysis.

Other than the generation technology, another key factor influencing the overall system strength is the network topology, i.e., how electrically far major generation and load centers are and how meshed and/or interconnected network areas are.

A power system with low system strength shows one or more of the following: larger voltage variations if power flows change, deeper voltage dips and higher over-voltages, prolonged voltage recovery after disturbances (emphasized in presence of induction motors), wider undamped voltage and power oscillations, malfunction (or failure) of protection equipment, increased harmonic distortion (to be considered especially in view of increasing RES penetration), commutation failure of LCC HVDC links, difficult synchronization, and negative impact on controls of some converters. In general, all these features clearly worsen the operation and security of electrical grids, but also the operation of converters specifically: in fact, system strength affects the ability of power system equipment to operate in a stable manner, but this holds true also regarding converter-based generation, which is ironically the main reason causing the fault level decay (together with the phasing out of thermal power plants).

Converter-based RES generation, as well as battery energy storage systems (BESS) and LCC HVDC links, require adequate system strength levels for their stable and reliable operation. This is because synchronous machines are electromagnetically coupled to the power system voltage waveform, whereas the RES generator is decoupled from the grid by the converter, and traditional GFL inverters do not create a voltage waveform like synchronous machines.

Complex interactions exist not only between a converter and the grid but also between connected inverters. GFL converters must synchronize with the grid voltage at their terminals and inject current at an angle that follows the measured voltage for their proper operation. By means of a phase-locked-loop (PLL), the inverter creates a synchronous clock driven by the voltage phase angle it senses from the grid. Following the occurrence and clearing of a fault, the inverter must re-lock onto the grid quickly to ensure stable control. Under low system strength conditions, the phase angle change between the pre-fault and the fault clearance is larger than on stronger systems, making the synchronization more difficult. If the voltage phase angle detected by a converter is inaccurate, the current is not injected correctly, impacting the voltage waveform to which it is connected and affecting in turn the voltage seen at its terminals. In strong systems, the voltage is marginally affected by the current injected by the converter, and the phenomenon is largely mitigated.

In an interconnected power system, these control interactions can have a cascading impact on the voltage waveform resulting in widespread disruptions if not corrected. The larger the number of converter-based RES, the greater system strength is needed at that location to maintain stability because there is a higher potential to influence the voltage. Hence, the ability of the network to withstand the change in voltage needs to be greater.

Therefore, it is crucial to increase system strength and upgrade converter technologies in the view of the power system weakening. Managing stability in low system strength conditions requires a combination of support from the network in conjunction with advanced and coordinated power electronic control.

9.4   Short-circuit analysis and converters

Traditionally, short-circuit computations have always been performed taking into consideration synchronous generators only and exploiting the superposition

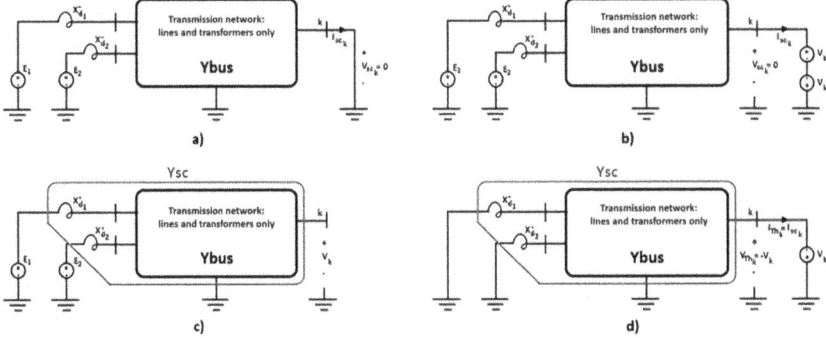

Figure 9.1   *Grid models under short-circuit: (a) overall model of the faulted grid; (b) overall model shown as a superposition of pre-fault and Thévenin state; (c) pre-fault state; (d) Thévenin state*

principle and the Thévenin theorem, taking advantage of their linear behavior under fault and modeling them as constant voltage sources in series to sub-transient reactances.

The fault is studied by superimposing pre-fault and Thévenin states, and the short-circuit is represented by two equal and opposite voltage sources, whose magnitude is the pre-fault voltage at that bus. Then, the circuit can be decomposed into two states being linear. The pre-fault state basically consists of the circuit corresponding to the normal operating condition before the fault, and the Thévenin one, made by the dead network and the negative voltage source on the faulted bus.

The approach is summarized in Figure 9.1.

Starting from short-circuit current, it can be stated:

$$E_{Th} = Z_{sc} \cdot I_{Th}$$

where E_{Th} and I_{Th} are the voltage and current vectors under the Thévenin state, respectively, and Z_{sc} is the impedance short-circuit matrix of the system (it can be derived as Y_{sc}^{-1}).

Assuming bus k as the faulted one and expressing the equation (fault current is exiting the bus, hence it is considered negative):

$$
\begin{bmatrix} E_{Th(1)} \\ \vdots \\ \vdots \\ E_{Th(k)} \\ \vdots \\ \vdots \\ E_{Th(n)} \end{bmatrix}
=
\begin{bmatrix}
Z_{sc(1,1)} & Z_{sc(1,2)} & \cdots & Z_{sc(1,k)} & \cdots & Z_{sc(1,n)} \\
Z_{sc(2,1)} & \cdots & \cdots & \cdots & \cdots & \cdots \\
\cdots & \cdots & \cdots & \cdots & \cdots & \cdots \\
Z_{sc(k,1)} & \cdots & \cdots & Z_{sc(k,k)} & \cdots & Z_{sc(k,n)} \\
\cdots & \cdots & \cdots & \cdots & \cdots & \cdots \\
Z_{sc(n,1)} & Z_{sc(n,2)} & \cdots & Z_{sc(n,k)} & \cdots & Z_{sc(n,n)}
\end{bmatrix}
\begin{bmatrix} 0 \\ \vdots \\ \vdots \\ -I_{sc(k)} \\ \vdots \\ \vdots \\ 0 \end{bmatrix}
$$

Therefore:

$$E_{Th(k)} = -Z_{sc(k,k)} \cdot I_{sc(k)}$$

Given that the Thévenin voltage magnitude at the faulted bus is equal to the pre-fault voltage:

$$I_{sc(k)} = \frac{E_{pre(k)}}{Z_{sc(k,k)}}$$

Accordingly, all bus voltages during the fault can be achieved by super-imposing the two states:

$$V_{fault} = E_{pre} + E_{Th}$$

Finally, the main quantity of interest related to system strength can be achieved: the short-circuit power of k-bus is equal to the product of its pre-fault voltage and the short-circuit current magnitude:

$$S_{sc(k)} = \left| E_{pre(k)} \right| \cdot \left| I_{sc(k)} \right|$$

The higher the short-circuit power, the higher the system strength, and the lower the equivalent impedance.

The contribution of synchronous machines is clearly derived by the presence, in the short-circuit impedance matrix, of their sub-transient impedances. The situation changes from what converters are concerned.

Concerning traditional GFL converter-based RES, their contribution used to be neglected, when their amount of generation was marginal compared to synchronous generators. This is no more acceptable in modern power systems. The non-linear behavior of converter-based RES generation implies that the current is bonded to the adopted control strategy and depends on terminal voltage, which is among the unknowns of fault analysis. Moreover, as extensively discussed in the previous section, PLLs need an external voltage source so as to keep the synchronism and drive the current, which is more difficult in case of low system strength; further-more, all converters in general must protect their valves, limiting the current and causing the aforementioned fault level decline, further emphasized by the phasing out of thermal power plants. These features are the reasons why voltage stability and system strength are crucial for GFL converters and their operation.

On the contrary, GFM is a recent technology whose electrical drive is inde-pendent of external voltage sources. GFMs are designed and can be controlled to emulate virtually synchronous machines, riding through faults, improving system strength, supporting the voltage in weak grids, and providing services such as fre-quency and voltage regulation, synthetic inertia, black start, and island operation. GFMs can generate their own voltage waveform (also under fault), imposing vol-tage and frequency on their own. In other words, their current is not driven, it is a consequence of load/fault, while voltage is controlled in magnitude and frequency and impedance is virtually emulated. GFM-based RES can operate properly in weak networks even supporting blackout restoration, in case of need.

It is clear, therefore, that all these features and constraints discussed so far should be considered by power engineers given their remarkable impact on fault current, to perform analyses properly and achieve reliable results in high-RES penetration power systems. Any converter-based source behaves according to its control, independent of physics, and every manufacturer implements its own controls; hence, it is difficult to compute exactly the contribution of each converter to the fault. That is why the approach proposed here is to start from the prescriptions given by Grid Codes, as will be discussed in the next section. The Grid Code rules, and in particular those related to the FRT capability, must be fulfilled by each converter, independent of the manufacturer. Therefore, the contribution of each converter-based RES was computed taking into account that common behavior.

Additionally, a couple of physical phenomena were considered not to neglect the impact of cascading events and to model grid condition during faults suitably, representing the impact of converters on short-circuit currents even more closely. First, RES disconnection in case their FRT limits are violated was modeled according to different grid code requirements. The second one is strictly focused on weak power systems: doubly-fed induction generators (DFIGs, Type 3) and Type 4 WTs are today the most widespread wind generation systems in weak grids, thanks to their high and decoupled powers controllability that allows mitigating voltage perturbations. Consequently, DFIG crowbar protection was considered for the proposed application, being another quite common element of this type of generation: it consists of a resistance inserted in series to the rotor in case the current reaches a threshold value as a consequence of excessive voltage drop, resulting in the bypass of converter and loss of rotor control capabilities [1].

9.5 Reference grid codes

Since converter current injections are strongly dependent on their control, whose peculiarities are typically patented by manufacturers and are usually not known, converter-based RES currents cannot be calculated uniquely, therefore have been assumed compliant with Grid Code's specifications. It makes sense to assume the converter control drive already operating in a sub-transient state given its speed.

In this chapter, reference will be made to two grid codes: the Italian Grid Code, issued by the Italian TSO Terna [13,14], and the German Grid Code [15,16], relevant to a power system where RES penetration is more remarkable; the latter is particularly interesting because of its prescriptions on the current to be injected in faulted conditions to support the voltage (Figure 9.2). Differently, the Italian code does not specify anything in this regard. By exploiting the algorithm, it will be possible to evaluate the positive effects of German Standards on the Italian scenario (whatever settings can be implemented depending on desired FRT requirements).

According to German code, RES currents must be in agreement with Figure 9.2: RES must provide a reactive injection proportional to the double of

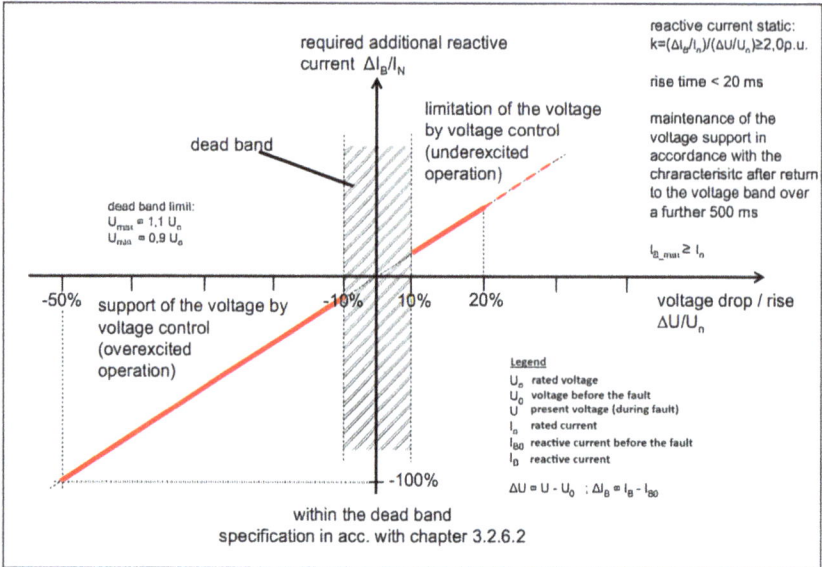

Figure 9.2   German FRT requirements

voltage variation (in per-unit). On the other hand, given the missing Italian requirements, all RES controlled according to Terna's Grid Code have been assumed injecting their pre-fault complex power until the inverter maximum current limits (assumed as 1.5 times the nominal one) is reached, otherwise the power is reduced at constant power factor keeping the current equal to its maximum.

9.6   Iterative short-circuit analysis

The core of the proposed algorithm is based on modeling the converter as a variable equivalent impedance draining the opposite of injected current during the fault, assuming the inverter current itself as a function of supplied power and bus voltage magnitude during the short circuit.

The procedure is as follows: after a preliminary power flow computation, standard short-circuit analysis is carried out neglecting inverters (i.e., considering them as infinite impedances), that means proceeding as discussed in Section 9.2. The resulting vector of short-circuit bus voltage magnitudes V_{sc} is used as the initial setpoint of an iterative procedure aimed at computing an equivalent shunt admittance of inverters y_{inv}:

$$y_{inv} = \frac{I_{inv}}{V_{sc}}$$

where I_{inv} are inverter currents related to V_{sc} in accordance with FRT requirements of the chosen Grid Code. Afterwards, equivalent admittances are added to short-circuit bus admittance matrix Y_{sc} on their corresponding position on the diagonal of the matrix. Then, the new matrix Y_{sc}' is used to perform another short-circuit computation, whose new voltage profile V_{sc}' allows in turn to achieve new currents:

$$I_{inv}' = y_{inv} \cdot V_{sc}'$$

Active and reactive current mismatches are computed, and if below the considered tolerance level $|I_{inv}' - I_{inv}| < \varepsilon$, it means that fault voltages of two iterations can be considered the same and convergence is reached. Otherwise, the procedure must be repeated until the convergence criterion constraint is fulfilled. The procedure is summarized in the flowchart of Figure 9.3 (where FRT current specifications are according to the German Grid Code).

RES disconnection and crowbar protection were implemented as follows: crowbarred WTs are modeled as voltage sources in series to a constant impedance added to Y_{sc}, excluding them from iterative computation according to their physical behavior (no longer non-linear being converter control lost). Instead, a more complex algorithm is needed for disconnection. A null converter current is obtained

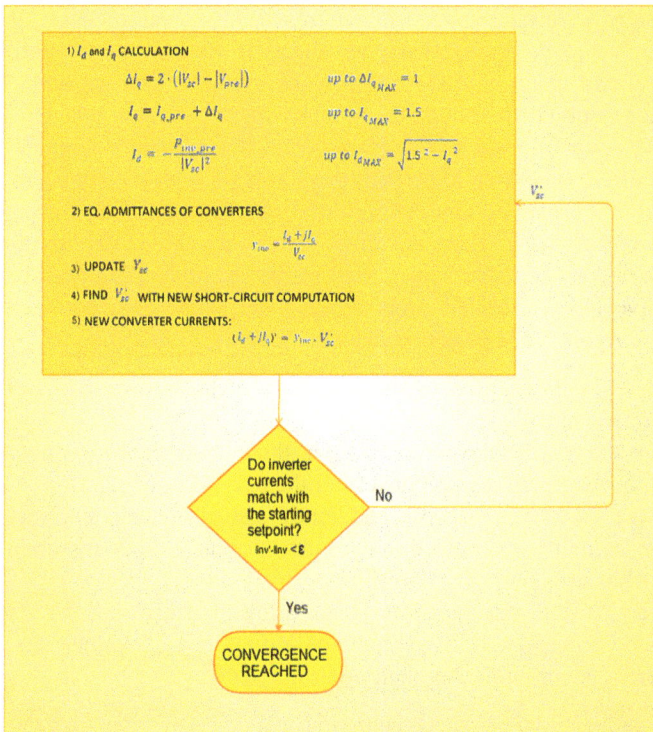

Figure 9.3   Iterative short-circuit approach including converters

with null equivalent admittance, but this is not correct physically: the disconnection is properly modeled if the overall current derived from a superposition of both pre-fault and Thévenin states is zero. Hence, another iterative loop was implemented under the Thévenin state, computing an equivalent admittance draining the opposite of the inverter pre-fault current. Consequently, superimposing the two states, the final current is zero, and disconnection is represented correctly.

Once convergence is reached, all voltages of current-controlled RES and crowbarred DFIGs are checked: if any voltage is lower than the defined threshold, the related generator is moved to the crowbarred WTs list or the disconnected RES list. Then, since RES lists were changed and the achieved solution is not the physical one, the short-circuit procedure must restart from the beginning. The selected RES generators are excluded from the iterative procedure and treated, in the next iteration, like explained above.

It is worth highlighting that, for most of the complex power systems, specific software packages of power system analysis can be used to model the network and run traditional short-circuit computation, implementing the iterative approach as an external code extension, starting from the commercial software output.

The approach has been extended to include also GFMs, according to their physical behavior [17–19], designing a new model of them focused on short-circuit analysis maintaining steady-state principles: in general, GFM can be modeled during faults as a constant voltage source in series to a reactance, i.e., its behavior can be assumed to be linear. However, since GFMs are power electronics converters, they must again limit their current according to their capacity. Therefore, the converter responds to voltage drops increasing its current, whereby exceeding its capability is not allowed by modifying voltage angle and magnitude.

If the current limitation is needed, there are two possible strategies available: either switching over to current control during faults to limit the current or implementing current clipping. There is still no commonly agreed strategy to overcome this issue in the power industry, and how current should be prioritized has not been defined univocally (if scaled proportionally or prioritizing active or reactive power). As a result, GFMs were modeled as follows:

- In general, GFMs are constant voltage sources in series to (virtual) sub-transient reactances.
- If GFMs exceed their overcurrent limit, they are switched to the current controlled mode, which means converting them to GFLs.
- Given the focus of this work and future issues of fault level decrease, the reactive current is prioritized under current control mode (assuming German Standards).
- In general, GFM disconnection is not allowed, since they have to emulate synchronous machines; it is accepted only if faults occur at their bus while they are current controlled being their control as GFL.

To sum up, both GFMs and synchronous machines are treated in the same way initially. Hence, sub-transient reactances are added to the admittance matrix and the Thévenin profile is achieved; then, it is exploited as an input for the iterative procedure where current controlled RES contributions are computed. Afterwards, once the fault voltages are obtained, GFM currents are checked: if any GFM exceeds its overcurrent limit, generator lists are updated and correspondent GFM is moved to the current controlled RES list. Then, since the solution is not the physical one, the procedure is restarted. The overall approach is summarized in Figure 9.4.

In conclusion, it is important to point out that the behavior of non-linear converters requires the evaluation of their currents iteratively, but once they are computed, RES can be treated as external ideal sources connected to a linear circuit/grid. In other words, the approach consists of a linearization valid about the operating point. Accordingly, the superposition principle, that is the key concept on which the analysis is based, is valid and can be adopted. Finally, when convergence is reached and modifications in RES lists no longer take place, short-circuit analysis is completed and achieved voltages and currents can be stored. An additional issue needs to be clarified: in the case of a direct connection of any RES generator

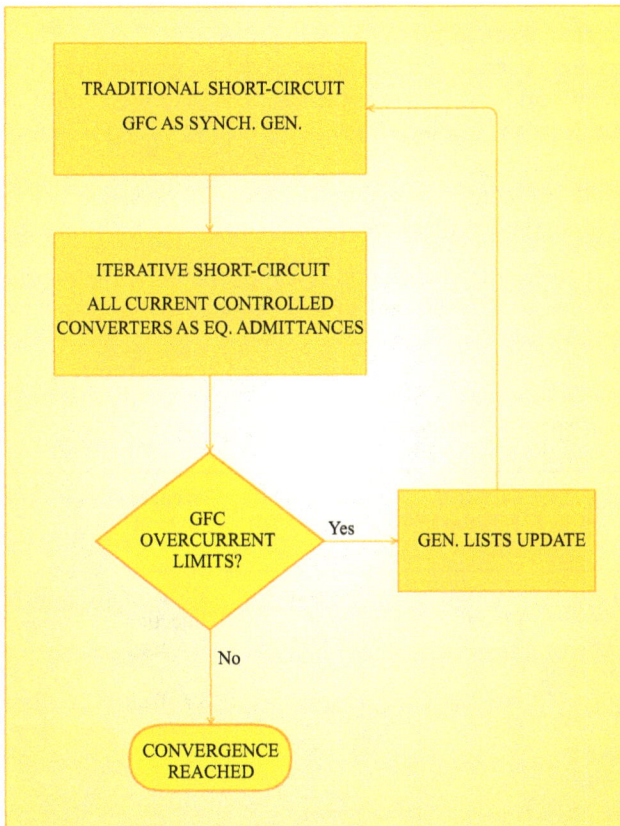

Figure 9.4   Iterative short-circuit approach including GFMs

to a faulted bus, it is suddenly moved to the disconnected RES list and excluded from iterative computation, in accordance with its physical behavior.

It is also worth underlying that the main aim of this analysis entails investigating the impact of RES on short-circuit power and system strength in view of their always higher penetration: hence, the tool has been designed focusing on three-phase short-circuit analysis modeling the grid with positive sequence only. Asymmetrical faults were not considered but could easily be implemented as discussed in [6], and electromagnetic phenomena were intentionally neglected because they are outside the scope of work. The result is a practical and reliable tool that represents the most significant phenomena and that perfectly suits operational planning and practical case studies.

9.7   The Sicilian grid: a real case study

The Italian power system is characterized by the largest RES converter-based generation located in the southern part of the peninsula, concurrently with low load and weaknesses due to few interconnections with the continental Italy, which is on the contrary more industrialized and interconnected with Europe. This often results in congestions and operational issues emphasized by intermittent RES generation. Hence, the Sicily's grid was chosen as the ideal context for investigating the impact of RES on fault level and testing different FRT control strategies; it is characterized by 181 RES out of 261 generators.

The network was contextualized in four different scenarios, and their features are summarized in Table 9.1 and presented below:

- Scenario 1: present-day Italian grid condition: no FRT requirements to support voltage, therefore RES are assumed to provide pre-fault complex power if

Table 9.1   *Summary of considered scenarios*

	Scenario 1 Terna 2021	Scenario 1.1 Terna 2010	Scenario 2 Hybrid	Scenario 3 E.On-VDE
HV RES control	Pre-fault power	Pre-fault power	Reactive injection	Reactive injection
HV DFIG crowbar	0.4 Vn	0.4 Vn	0.4 Vn	–
HV WT disconnection	–	0.2 Vn	–	–
MV RES control	Pre-fault power	Pre-fault power	Pre-fault power	Pre-fault power
MV DFIG crowbar	–	–	–	–
MV WT disconnection	–	–	–	–

below current limits. No disconnection is allowed, but HV DFIG crowbar protection is accepted if the bus voltage is 0.4.

- Scenario 1.1: Italian grid condition of 10 years ago: it was considered to evaluate the direction in which improvements were oriented and to investigate Grid Code evolution results; it is almost the same as Scenario 1, except HV WT disconnection allowed if bus voltage is 0.2.
- Scenario 2: a hybrid scenario based on nowadays grid except for HV RES reactive injections which are carried out in accordance with German Grid Code.
- Scenario 3: it is based on all German requirements: reactive injections under fault and no crowbar protection allowed. Reactive controllability must never be lost.

In all scenarios, RES disconnection is allowed if the fault is at their terminals. MV RES are always controlled according to Italian Standard CEI 0-16 [20], i.e., injecting pre-fault complex power since again no FRT requirements are specified to support the voltage.

9.8   Procedure testing and dynamic simulations

Before starting with short-circuit studies, the iterative approach is tested to prove its validity. A first comparison of results achieved with traditional and iterative methods was made considering Scenario 1. As shown in Figure 9.5, the percentage

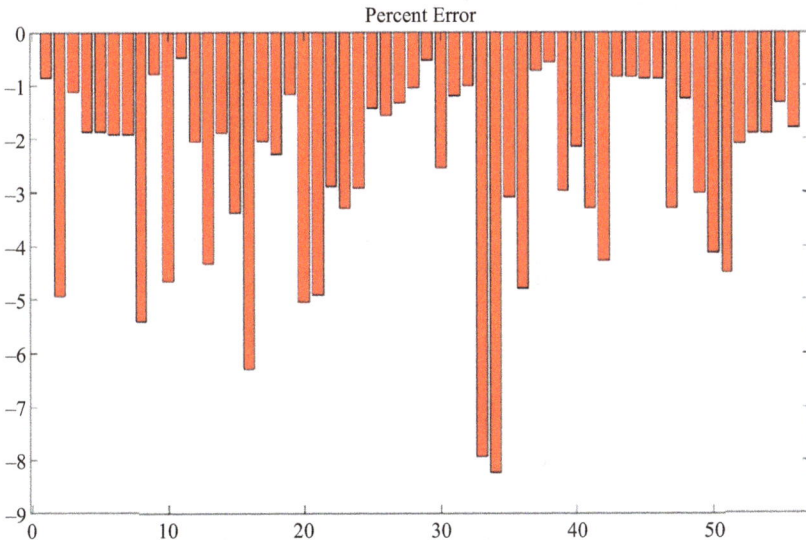

Figure 9.5   Short-circuit power errors of the 56 EHV buses if converters are neglected [%]

error neglecting RES is not negligible being up to 9%, too high to accept the approximation.

It is worth mentioning that convergence could result in difficult to reach with the increment of RES penetration, being the analysis based on Gauss iterative method.

Hence, iterations were limited to 15 with a tolerance threshold of 5e–4 p.u., both on active and reactive currents. If 15 iterations were not enough, the iterative computation was decelerated adding 25 more iterations using at each one the average of the actual short-circuit voltage value and the previous iteration one.

Furthermore, to check the accuracy of the proposed iterative approach, several dynamic RMS simulations have been performed, modeling the Sicilian power system in PSS/E: synchronous generators (machines, turbine governors, and exciters) and RES generators (converters and converter controllers) were modeled with standard PSS/E library models. To compare results more consistently with PSS/E, RES converters were set injecting their pre-fault complex power and neither crowbar protection nor disconnection were allowed. Hence, starting from typical fault analysis assumptions, such as neglecting shunt admittances and loads, three-phase short-circuits at the most important buses were simulated. Voltages are focused on, being the main unknowns and given that converter injections depend on them: in fact, if voltages are the same for the iterative method and dynamic simulations, not only the quality of the approach is good but also RES are proven to inject the same currents in both static and dynamic studies.

For instance, voltages of some significant buses are reported in Figure 9.6 with a three-phase short-circuit occurring at bus 538 (the bus which presents the highest short-circuit power of the system): as can be appreciated, differences are negligible, confirming the validity of the iterative approach. The other simulations confirmed these results.

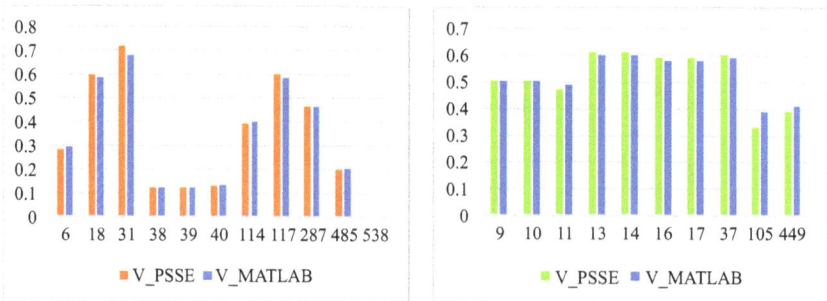

Figure 9.6   *Three-phase short-circuit at bus 538; from left to right: voltages of main EHV buses, and voltages of some relevant HV and MV RES buses*

9.9 Fault level of Sicilian power system

Short-circuit powers of EHV buses in Scenarios 1, 2, and 3 are reported in Figure 9.7 (left-hand). As shown, the more RES reactive support increases, the higher the short-circuit power. These results show the benefits on power system strength that a similar upgrade of the Italian Grid Code would imply, especially in view of future RES penetration. Since differences between the second and third scenarios are almost negligible, an upgrade limited to Scenario 2 would be enough to obtain significant results. In addition, higher fault level and better voltage support would mitigate troubles related to protections and fault detection, and improve angular stability, crucial in the presence of long backbones as in the case of Southern Italy. Concerning Scenario 1.1 (the former Italian scenario), as shown in Figure 9.7 (right-hand), no valuable improvements are reached foreclosing disconnection of WTs: increments are infinitesimal and not even noticeable by the chart. This suggests that preventing RES from disconnection allows the avoidance of a no longer acceptable generation loss, but it does not affect short-circuit power.

9.10 Fault level of 100% RES power system

A hypothetical power system was modeled to investigate future system strength conditions in 100% RES-based grids: all synchronous generators of Sicily's network were substituted with RES based on GFL converters, except the slack bus of the system model, located in the Calabrian power station of Rizziconi, which had been modeled to also represent the impact of the peninsular grid (thereby enabling the operation of GFLs physically). Given the nature of the scenario, German FRT Standards of Scenario 3 were assumed as RES control mode.

The comparison of the current grid fault level and the hypothetic grid one is reported in Figure 9.8 (blue and yellow charts, respectively): short-circuit power reduction is dramatic, especially noting that EHV buses are considered.

Looking for improvements, the German rules were extended to all RES including MV ones. As shown in Figure 9.8 (orange chart), the enhancement is significant, confirming once more the important impact of this strategy on power system strength.

These results allow getting an idea not only about the decadence that fault level would experience in case of 100% RES penetration but also how RES reactive injections under fault can contribute substantially to improve the operating condition of a similar scenario.

9.11 Comparison of grid forming and grid following operation

Finally, in searching for additional operating enhancements and to investigate their impact on system strength, GFMs have been considered in Scenario 3 and prioritizing reactive injections. Computations with GFMs were carried out both in the

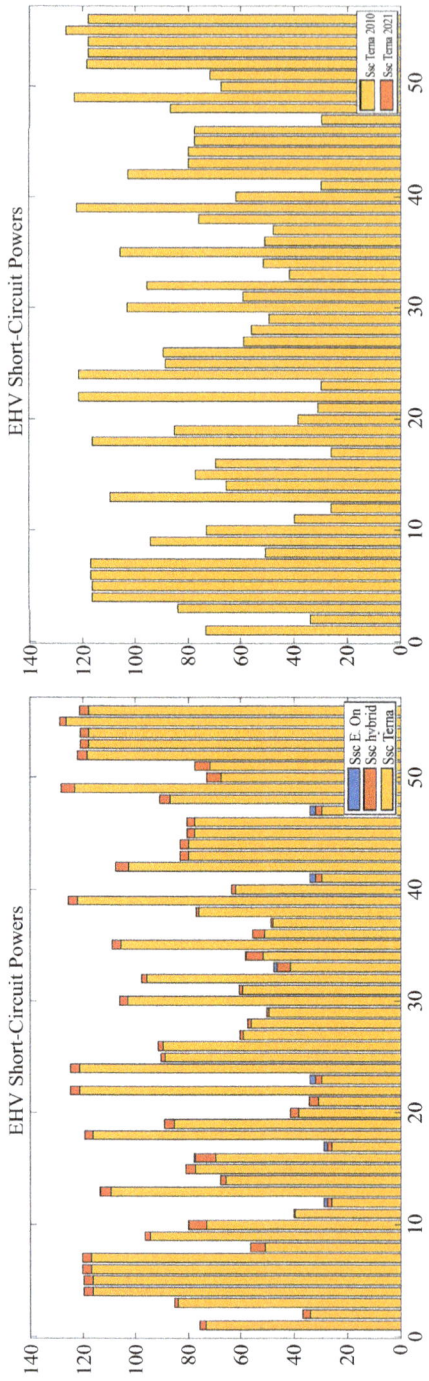

Figure 9.7 EHV buses fault level of Sicilian grid in per-unit (100 MVA)

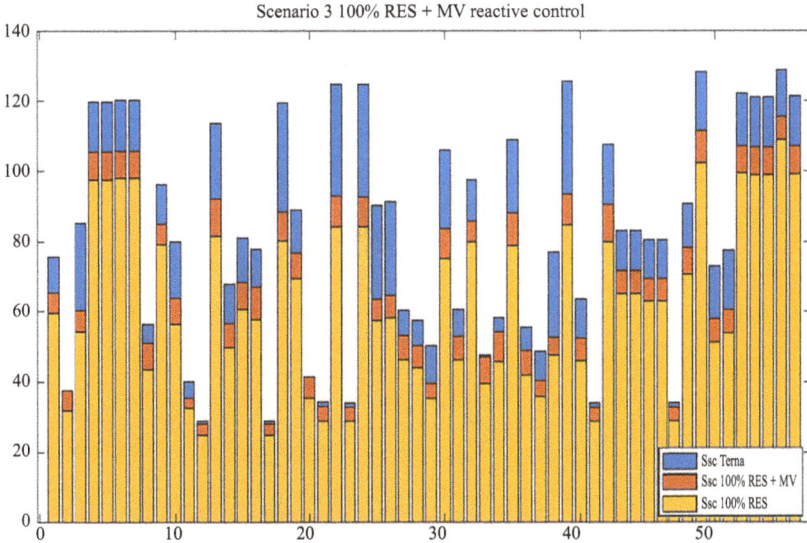

Figure 9.8   *EHV buses fault level with different scenarios and reactive support in per-unit (100 MVA)*

future 100% RES scenario as well as in the present-day one, replacing all traditional GFL converters with GFMs. The comparison between two converter technologies in both scenarios is reported in Figure 9.9: in 100% RES grid (left-hand), GFMs improve significantly fault levels in almost all buses. In some cases, short-circuit power is the same since GFMs reaching their current limits behave as current-controlled converters.

The same does not hold for present-day scenario (right-hand): short-circuit powers are almost the same, except for a slight deterioration in the presence of GFMs. This can be physically motivated by recalling very recent studies, in which a duality between GFM and GFL is discussed [21,22]: analyzing inverter control structures and operation, GFMs were proven to work better in low fault level and high impedance grids, differently from well-known GFLs, whose PLL control requires the opposite condition. Hence, grid strength plays a relevant role in GFM operation as well.

In light of this, computation results can be motivated as follows: given the fault level drop in 100% RES grid, GFMs provide their positive contribution to short-circuit power; on the other hand, in the present-day scenario (with the German Code), the fault level is higher being sustained by synchronous machines and, despite its weakness, the Sicilian grid is still a meshed power system with relatively low equivalent impedance. Hence, GFMs can be used, but they are not in their ideal operating context. Accordingly, despite reactive injections, short-circuit power presents a slight decrease, even if limited given the rotating generators. In other words, grid strength does not result in a bad condition for GFMs, but at the same time, their performance is worse than the GFLs supporting the voltage.

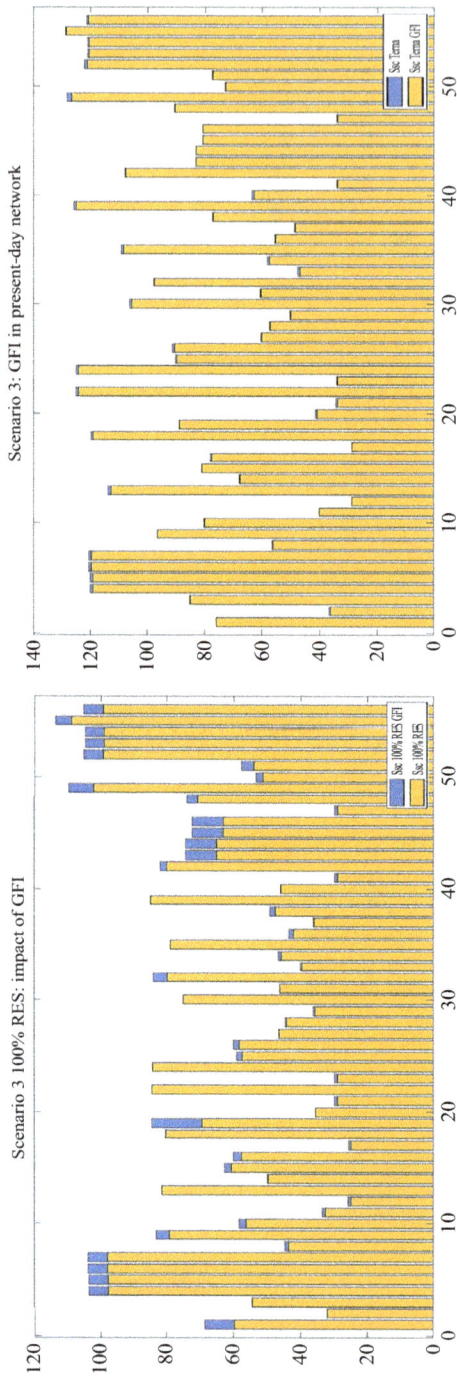

Figure 9.9 EHV buses fault level in 100% RES and the present-day scenarios with RES based on both GFL and GFM converters

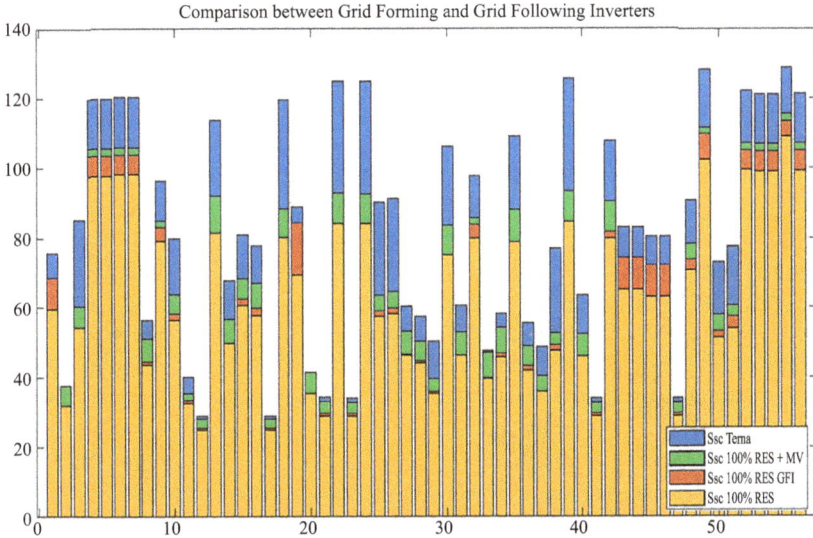

Figure 9.10   EHV buses fault level in 100% RES grid with different converter control strategies in per-unit (100 MVA)

Therefore, a final computation was carried out in 100% RES scenario, demonstrating, as reported in Figure 9.10 (based on Figure 9.8), that a massive reactive compensation carried out by all RES (both HV and MV ones, green chart) connected via GFL provides better enhancements rather than connected via GFM (orange chart). These results confirm what was stated above regarding better short-circuit contribution of GFL in meshed power systems and GFM in low fault level networks, suggesting that in the future the best strategy would involve the cooperation of both converter technologies depending on the grid strength instead of focusing on GFM development only.

9.12   Future research

This chapter has presented a methodology that, keeping the simplicity and pragmatism of the approach typical of Standards, tries to address a very complex problem related to a branch of the power system strength. However, it is clear that in the next future, given that power systems in the next decades will be characterized by parameters completely different from today's, new methods and tools will be needed.

One research direction very clearly emerging from the technical reports dealing with converter-based generation is that weak systems will be more and more common. Today, power system strength issues are reported in a few countries, but the first scenario analyses carried out for systems which are nowadays very stiff show that they also will soon suffer from very low short-circuit levels, becoming in turn weak. The industry is aware of the risk of being not prepared for that situation and both transmission and distribution system operators on the one side, and manufacturers on the other side, are dealing with this issue, trying to fill that gap.

The main research trends can be resumed as follows: low power system strength values result in an increasing difficulty for the synchronization of power converters, especially for GFL converters, whose PLL can be negatively affected by the output current of the converter; the compliance with FRT curve established in different countries, and the possibility to change the reactive current injected by the converter during a fault; the transition between grid-isolated and grid-connected operating conditions, that can start oscillations to be stabilized; the possibility of converter-based generation to provide voltage and frequency control services, in different operating conditions (e.g., PV converters during night periods).

All the above-mentioned subjects, related to both small-perturbation and large-perturbation stability, together with different control schemes for the real and reactive power are affected by the short-circuit level and stiffness of the network. To cope with these subjects, however, it is necessary to investigate the controls implemented by manufacturers, which are usually patented, not publicly available and not standardized. Here an issue about how to evaluate interactions among different converters comes in, and it is not of easy solution. Actually, a common conclusion of most recent research papers is that steady-state and RMS simulations are no more accurate enough to capture all instability issues and all interactions among converters, in weak systems. EMT simulations are becoming necessary to model all machines and interaction phenomena; such models are today usually not available, and, in particular, they are not public, making it difficult to derive general conclusions on possible negative effects and to set up standard procedures.

Manufacturers should provide system operators (at least) with black-box models of their controls and machines for EMT simulations to be carried out on the actual model of the power system to be investigated. On the other side, systems operators should arrange suitable computational resources able to tackle large problems in the EMT environment. Moreover, human resources specifically trained for that should be dedicated to the task: in facts, the competences of such power engineers will be in the field of power systems, power converters, controls, and communications, i.e., very broad compared to past system engineers.

Together with the research work above mentioned, from the industrial point of view, it is more important to be able to translate all research outcomes into Standards, as simple as possible (not an easy task, in this case!), and prototyping procedures. The present chapter presented a methodology suitable for the nowadays technical scenario, but in the next future, it will be necessary to move on and provide even more accurate procedures, able to capture more complex interactions among converters.

9.13   Conclusions

Given the no longer negligible impact of RES, a new approach to short-circuit analysis has been proposed to keep into account the contributions of both GFL and GFM converter technologies, resulting in a complete and effective tool that makes it possible to correctly compute fault currents and to include cascading events without resorting to dynamic simulations. The algorithm was tested and exploited to carry out operational planning studies in different scenarios so as to investigate the impact of RES penetration on short-circuit and system strength. Several FRT strategies were analyzed, and the update of the Italian Grid Code was discussed, in search of future operational improvements. All analyses carried out have shown, as a minimum common denominator, how beneficial it would be for the Italian power system to update the Grid Code following the direction of German Standards. Moreover, both converter technologies and their impact on short-circuit were studied, comparing GFM and GFL to highlight their peculiarities and benefits on system strength to define their future role in electrical grids, looking forward to an energetic transition and a completely decarbonized power system.

Bibliography

[1] R. A. Walling, E. Gursoy, and B. English, Current contribution from Type 3 and Type 4 wind turbine generators during faults, In *IEEE/PES Transmission and Distribution Conference and Exposition (T&D)*, 2012.

[2] V. Helac and S. Hanjalic, Wind farm response on short circuits and longitudinal asymmetries, In *19th International Symposium INFOTEH-JAHORINA*, 2020.

[3] N. Leonov, G. C. Cho, and A. Poluektov, Short-circuit currents from wind turbine generators study in laboratory workshop, In *V International Conference on Information Technologies in Engineering Education*, 2020.

[4] N. Berisha, P. Kastrati, D. Stoilov, and R. Stanev, Analysis of short-circuit currents in interface nodes of Kitka Wind Power Park in Kosovo, In *2021 17th Conference on Electrical Machines, Drives and Power Systems (ELMA)*, 2021.

[5] E. Muljadi, M. Singh, R. Bravo, and V. Gevorgian, *Dynamic Model Validation of PV Inverters under Short-Circuit Conditions*, IEEE GreenTech, 2013.

[6] Ö. Göksu, R. Teodorescu, B. Bak-Jensen, F. Iov, and P. C. Kjær, An iterative approach for symmetrical and asymmetrical short-circuit calculations with converter-based connected renewable energy sources. application to wind power, In *IEEE Power and Energy Society General Meeting*, 2012.

[7] H. Margossian, J. Sachau, and G. Deconinck, Short circuit calculation in networks with a high share of inverter based distributed generation, In *IEEE 5th International Symposium on Power Electronics for Distributed Generation Systems (PEDG)*, 2014.

[8] D. Duckwitz, A. Knobloch, F. Welck, T. Becker, C. Glöcker, and T. Buelo, Experimental short-circuit testing of grid-forming inverters in microgrid and interconnected mode, In *NEIS, Conference on Sustainable Energy Supply and Storage Systems*, 2018.

[9]   M. Nahidul, K. Meng, W. Xiao, A. Al-Durra, and Z. Y. Dong, Interactive grid synchronization-based virtual synchronous generator control scheme on weak grid integration, In *IEEE Transactions on Smart Grid*, 2021.

[10]   Y. Li, K. Meng, and Z. Y. Dong, Frequency enhancement of grid-forming inverters under low-SCR weak grid, In *8th Renewable Power Generation Conference (RPG)*, 2019.

[11]   Australian Energy Market Operator (AEMO), *System Strength*, March 2020.

[12]   B. Badrzadeh, *System Strength*, CIGRE, ELECTRA no. 315, April 2021.

[13]   Attachment A68 of Terna's Grid Code, *CENTRALI FOTOVOLTAICHE-Condizioni generali di connessione alle reti AT, sistemi di protezione, regolazione, controllo,* 2019.

[14]   Attachment A17 of Terna's Grid Code, *CENTRALI EOLICHE-Condizioni generali di connessione alle reti AT, sistemi di protezione, regolazione, controllo*, 2019.

[15]   On Netz GmbH, *Grid Code for High and Extra High Voltage*, 2006.

[16]   VDE, VDE-AR-N4120, New German Transmission Code for onshore connections at 110 kV (or above), 2017.

[17]   B. Badrzadeh, Z. Emin, S. Goyal, *et al.*, *System Strength*, published in "Towards System Strength" by CIGRE Science & Engineering, Innovation in the Power Systems Industry, vol. 20, February 2021.

[18]   P. Christensen, G. K. Andersen, M. Seidel, *et al.*, *High Penetration of Power Electronic Interfaced Power Sources and Potential Contribution of Grid Forming Converters*, ENTSO-E, 2020.

[19]   Y. Lin, J. H. Eto, B. B. Johnson, *et al.*, *Research Roadmap on Grid-Forming Inverters*, NREL, 2020.

[20]   CEI 0-16, *Reference Technical Rules for the Connection of Active and Passive Consumers to the HV and MV Electrical Networks of Distribution Company*, 2019.

[21]   Y. Li, Y. Gu, and T. C. Green, *Rethinking Grid-Forming and Grid-Following Inverters: A Duality Theory*, IEEE, Jun 24, 2021.

[22]   High Share of Inverter-Based Generation Task Force, *Grid-Forming Technology in Energy Systems Integration*, Energy Systems Integration Group (ESIG), March 2022.

[23]   L. Vignali, *The Impact of Renewable Energy on Voltage Stability and Fault Level in Power Systems,* Master Thesis, Politecnico di Milano, AY 2020-2021.

Chapter 10

New smart devices-based strategies for optimal planning and operation of active electric distribution networks

Gheorghe Grigoras[1], Bogdan-Constantin Neagu[1], Livia Noroc[1] and Ecaterina Chelaru[1]

Abstract

The rapid evolution of the passive distribution technical infrastructure towards one active was due to the emergence of new innovative technologies that improved its efficiency and reliability. These include the use of smart meters and automation devices. The increasing number of small-size renewable energy sources (RESs) has represented another challenge to the policy of the distribution network operators (DNOs) to register the direction of the target regarding the minimization of greenhouse gas emissions associated with the distribution process. Numerous studies revealed that an increasing number of end-users will want to become active in the electricity sector in the next period, which can have significant implications for the system strength. The equipment manufacturers developed innovative solutions based on smart meters and other advanced devices in the active electric distribution networks (AEDNs) to provide reliable and flexible communication systems. The availability of dedicated platforms and modern technologies allowed the identification of new solutions, with a high impact on the system strength. Based on these technological evolutions, the DNOs must develop efficient strategies to improve the operating conditions of the AEDNs and reduce the environmental impact based on innovative technologies to integrate advanced techniques and real-time communication solutions.

In this chapter, the authors proposed a new smart devices-based strategy in the optimal planning and operation of the AEDNs integrating the small-size local renewable generation sources with various penetration degrees to improve the system strength. Testing has been done in a Romanian AEDN, and the results confirmed obtaining the technical and economic benefits through its implementation.

[1]Electrical Engineering Faculty, "Gheorghe Asachi" Technical University of Iasi, Romania

10.1   Introduction

10.1.1   *The AEDN concept*

The European Union (EU) is engaged in the goal of the Paris Agreement regarding climate change and is planning on becoming climate neutral by 2050 [1]. As part of the Green Deal, the European Union agreed in 2021 to reduce greenhouse gas emissions by 55% by 2030. This target, which is an interim one, will be used to develop more detailed policies and targets in the coming years. In 2020, emissions in the EU were 33% below 1990 levels [2]. Regarding the invasion of Ukraine, the need to enhance the EU's security and reduce its dependence on Russian fuels has been highlighted. In this economic and geopolitical context, the problem arises that the technical and communication infrastructure of the European power system to be as soon as possible modernized to fulfil these targets, and one of the objectives of this process is represented by accelerating the transition towards AEDNs. The traditional electric distribution networks (EDNs) should integrate into a percentage as high as possible the RESs and the technological advancements that have occurred in the field. The shift toward more flexible and efficient EDNs is happening more and more quickly due to the increasing number of distribution network operators (DNOs) looking to invest in replacing the passive architecture with an active one [3,4].

An AEDN is a new concept that integrates renewable energy resources (RESs), energy storage systems (ESSs), and controllable loads (electric vehicle charging stations, prosumers with battery energy storage systems, prosumers/consumers with electric vehicles) in the distribution infrastructure, see Figure 10.1. Also, it is associated with a digital "revolution" which allows an increase in the capacity to process and transmit the significant data amount using the advanced communication information technologies [4,5].

Today, equipment producers offer a diverse spectrum of innovative products and advanced solutions for distribution automation and the communication networks included in the AEDNs, covering a wide range of data analysis techniques based on digital solutions and Internet of Things (IoT) applications. The rollout and integration of innovative solutions also involve a fundamental repositioning of distribution services, although the typical use of the term focuses on technical infrastructure. The emergence of modern technologies has allowed DNOs to design, plan, and operate it reliably [4,6].

The rapid development of AEDNs has shifted the energy distribution sector toward one more service-oriented. The efficiency of this process can be improved using the smart metering system (SMS) integrated between the network and end-users. The SMS represents a critical piece of the informational infrastructure, enabling the integration of hardware, software, and data management. The AEDNs can be managed effectively with the help of the smart meters assimilated by the DNOs with the heart of the AEDNs. Also, smart meters could be considered the gateway to new communication and data technologies being able to offer innovative services to all end-users. A study conducted in December 2019 revealed that by

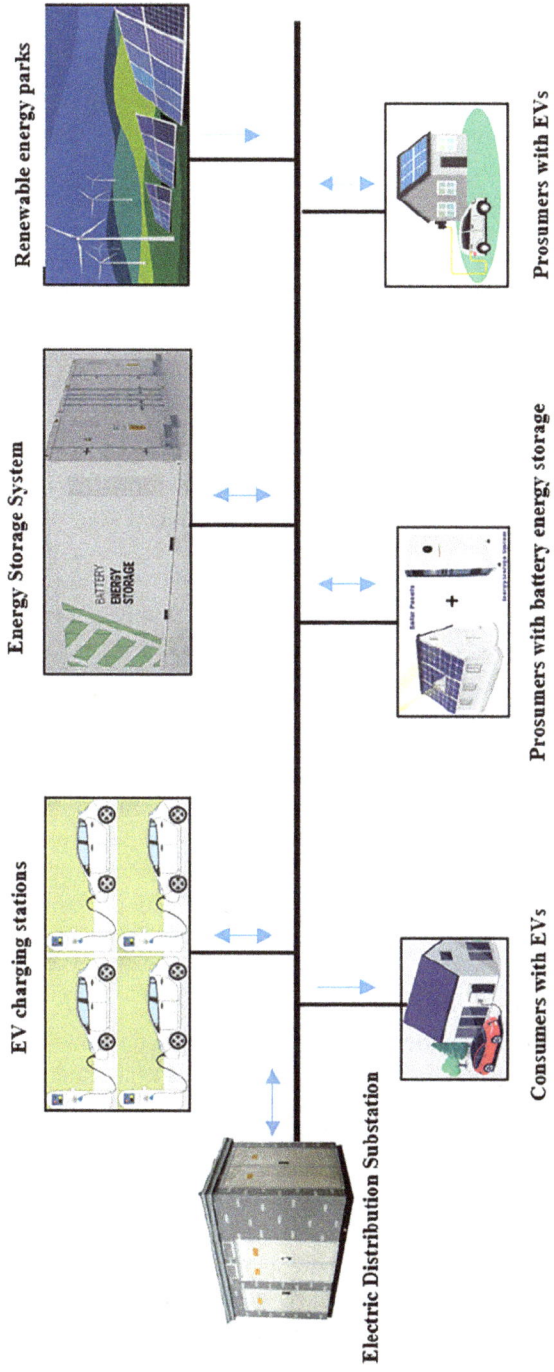

Renewable energy parks

Energy Storage System

BATTERY ENERGY STORAGE

EV charging stations

Electric Distribution Substation

Prosumers with EVs

Prosumers with battery energy storage

Consumers with EVs

Figure 10.1   The components of an AEDN

2024, over 200 million smart meters are planned to replace the old meters in the EU, and more than 70% of consumers will have a smart meter for electricity [7]. The smart meters can provide savings of up to 270 euros for electricity per meter point, according to data collected from pilot projects [8].

Through the use of smart meters, it can establish a development framework of new services and solutions to help them manage the energy consumption of the end-users receiving timely updates on the electricity tariff. Thus, the decisions of electricity-intensive industries and citizen energy communities will consider the changes in the energy price. In this context, the new market players will develop new models and solutions that will help them meet their customers' needs, which will also contribute to the emergence of a more competitive retail market and the reduction of greenhouse gas emissions.

The various electric mobility and demand response solutions can represent the advantages for the end-users and energy communities with smart communication infrastructures available to decrease the cost of electricity bills. Finally, all actors (DNOs, suppliers, end-users, and producers) will quantify the benefits of new technologies integrated into the smart meters. On the other hand, the increased connection capability of decentralized small-size RESs (mainly PV systems) in the AEDNs will represent another challenge for DNOs to predict their operation and respond to disturbances efficiently [9]. In addition, the integrated software solutions will help to support the digitization of the power supply and an optimal steady state of the AEDNs, improving the operation of the RESs [7].

Distributed PV generation has become a worldwide concern regardless of the country. Although the penetration degrees at the level of the whole distribution system did not have a significant increase, however, elevated penetration levels in certain areas on an increasingly large number of distribution feeders and the installation of the smart meters at all end-users have opened the gate to the rapid transition towards to AEDNs. On the other hand, these decentralized small-size RESs can help reduce the environmental impact of electricity production, but they can also include risks due to the uncertainty involved in implementing them [10,11].

10.1.2   *The basics of power systems strength*

The term "strength" represents a fundamental concept in the power system used to measure how well the system interacts with the devices connected to it regarding a load, a power generation unit, or an electric station.

According to Urdal *et al.* [12] and based on the [13,14], the power system strength can be defined from two aspects: the first is represented by its impedance which is made up of the components (generators, transformers, transmission lines, and loads), and the second corresponding to its mechanical rotating inertia. If the value of the impedance is high, system strength is weaker, leading to voltage variations and other undesirable technical issues. Regarding the second aspect, mechanical rotating inertia should ensure a stable frequency and fix voltage angle variations.

Xu *et al.* [15] have explained the power grid strength based on the classic concept associated with the infinite power source having two characteristics: The first is that the voltage amplitude at each bus will not change regardless of the type or capacity of the device connected to it, and the second is that the frequency of the voltage at each bus will remain constant. The Australian Energy Market Operator regards the system strength from the perspective of the power system stability under all reasonably feasible operating conditions [16]. However, it is not necessary to consider system strength to ensure that the system is secure in the case of the classical power system without RESs. But, the rapid emergence and growth of RESs, such as wind and photovoltaic power, have increased their number in many countries from all continents. The RESs having high installed capacities are usually located in remote regions and connect to the power system through long-distance transmission lines. In these conditions, the equivalent impedance can have a high value leading to a weak power system. But, the impedance can change due to various factors, such as steady-state and fault regimes, even if the power system is considered strong.

The system strength should be evaluated before promoting the widespread adoption of RES technologies identifying if these can be critical since their connection to the power system can pose a significant challenge [17].

More studies indicated that RES technologies could influence system strength. One of these is associated with synchronous machines. Although inverter-based resources have been seen in the first phase as a negative factor on system strength, the emergence of new technologies, such as advanced grid-following inverters and grid-forming inverters, is starting to make a positive impact [18,19].

Regarding the AEDNs, DSOs have been able until now to integrate a growing more and more number of small-size distributed generation sources due to the AEDN strength and reasonable installed capacities. But, higher growth of these sources will represent a challenge for the DSOs, associated with their ability to maintain the network operating smoothly without technical issues [6]. The strength of a power system integrating large-scale solar or wind power plants is different from that of a distributed system with many PV prosumers integrated. These PV-rich distribution networks are more prone to issues regarding voltage quality, phase loading unbalances, or distribution capacity [20].

10.1.3   Original contributions

Due to the above-highlighted technical issues, efficient operation and planning strategies should be initiated by the DNOs considering the innovative technologies which allowed the development of smart devices such that the AEDN strength is ensured. Through their integration into the electricity distribution process, the DNOs could identify the best solutions regarding the optimal operation and planning of AEDNs.

Thus, a strategy based on the data provided by the SMs integrated at the level of the end-users (consumers and prosumers) and the control of the PLB smart devices, installed to the prosumers having single phase (1-P) branching connected to one of the three phases, and the on-load tap changer (OLTC) of the transformer from the EDS, has been proposed and analyzed in this chapter such that the AEDN strength to increase. Through its implementation, the optimal solutions regarding the operation and planning of the AEDN can be identified, starting from the current situation (topology and the end-users connected at the phases) and developing scenarios associated with the increasing trend in the number of prosumers quantified through a penetration degree, calculated as the ratio between the number of prosumers and the total number of the end-users.

The rest of the chapter is structured as follows: Section 10.2 presents the technologies integrated into the AEDNs, Section 10.3 explains the implementation stages of the proposed strategy, Section 10.4 presents a case study with details on the obtained results, Section 10.5 presents the research gaps, challenges, and future research directions, and Section 10.6 includes conclusions.

10.2   Technologies integrated into the AEDNs

Since the beginning of the 21st century, opportunities have become apparent to take advantage of the evolution of information technologies to solve the limitations and costs of AEDNs. The success of any initiative that involves the transition towards the AEDNs depends mainly on the integration degree of the smart devices and technologies. The operation and planning activities of the AEDNs carried out by the DNOs are focused on solving the issues related to the electricity market and the requirements of its end-users [4,21]. These development directions are designed to respond to these issues by developing strategies that address them by combining the communication infrastructure with software elements, control equipment, and smart devices. Thus, the distribution automation process started in most countries, and for others, it has ended, being the first step of the transition from passive to active EDNs, influencing the AEDN strength.

In this context, distribution automation based on innovative technologies integrated into an AEDN can respond rapidly to changes in the electricity demands of the end users. The technologies, applications, benefits, and research and development directions associated with the integration level (hardware, distributed monitoring and control, network operation based on the procedures and strategies) from an AEDN, according to [3], are presented in Figures 10.2–10.4. Integration of modern technologies can lead to the improvement of the AEDN strength.

Increasing the market of automation devices/equipment and the widespread installation of smart meters enabled significant absorption of technologies based on Artificial Intelligence techniques and the IoT, significantly influencing the operation and planning of the AEDNs. On the other hand, the digitization process and advanced cloud computing technology will contribute to monitoring consumption/ generation patterns and providing real-time information [22].

The new technologies integrated at the level of consumer metering have reduced the pressure on the estimation of the state of the networks and the reduction of energy losses (especially those associated with the commercial component). In

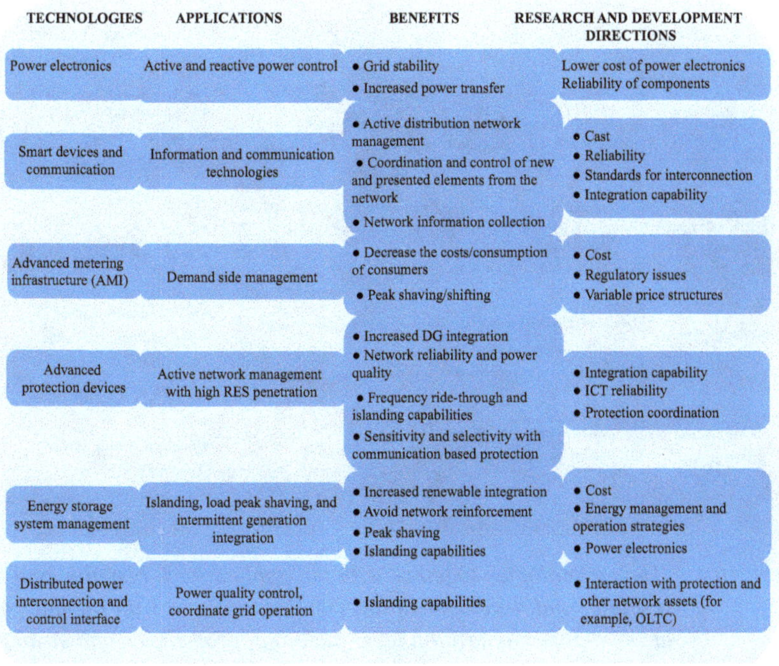

Figure 10.2   *The technologies together with the applications, benefits, and the research and development directions associated with the integration at the hardware level, according to [3]*

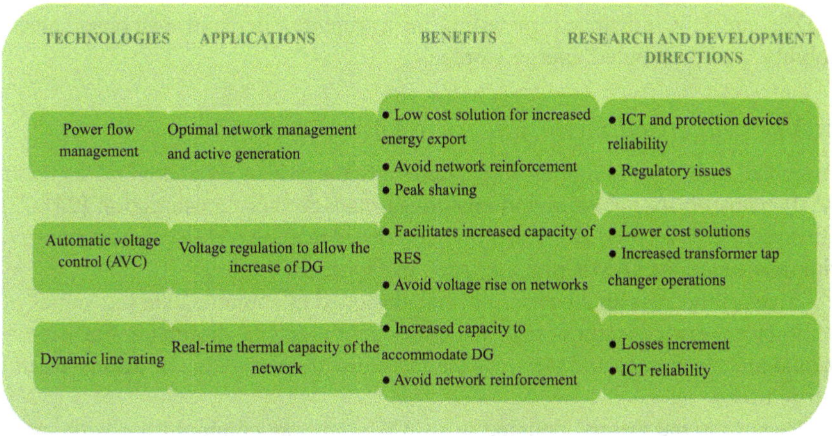

Figure 10.3   *The technologies together with the applications, benefits, and the research and development directions associated with the integration at the distributed monitoring and control level, according to [3]*

TECHNOLOGIES	APPLICATIONS	BENEFITS	RESEARCH AND DEVELOPMENT DIRECTIONS
Demand side management	Load reduction / shifting, price responsive, direct load control	• Frequency control • Reduce consumer costs • Avoid network reinforcement • Peak shaving	• Customers behavior • Regulatory issues • Variable price structures
Distributed control of RES	Aggregation of small and medium RES	• Integration of DG to optimal operation and economic maximization • Balancing of variable generation	• Cost • Regulatory issues
Microgrid / feeder (islanding)	Operation of small communities and buildings or a feeder of a substation in islanded mode	• Autonomous power supply • Avoid network reinforcement • Lower network losses	• Cost of energy storage • Resynchronization with the main grid • Protection schemes
Distribution management systems	Active management of distribution networks with RES integrated	• SCADA • Optimal power flow • Integration with IEDs and RTUs • Web access	• Cost • Integration with systems already in use • ICT reliability

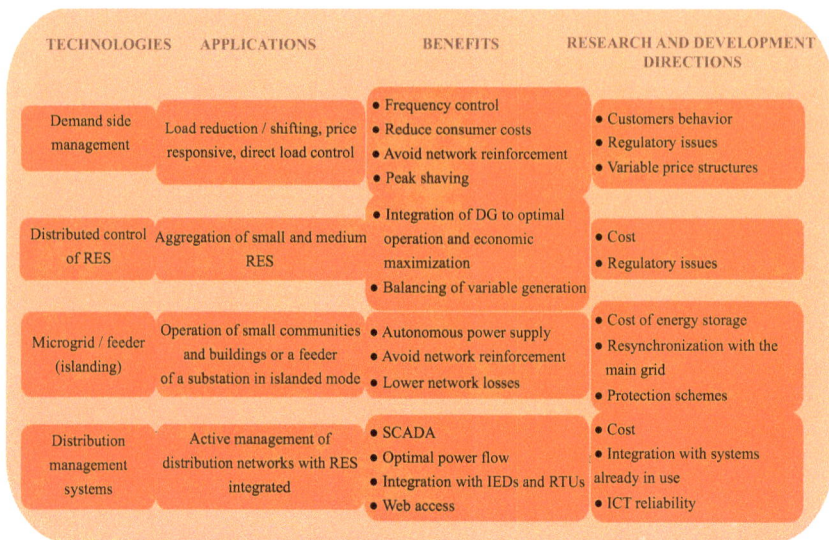

Figure 10.4   The technologies together with the applications, benefits, and the research and development directions associated with the integration at the network operation level including procedures and strategies, according to [3]

parallel, growing concerns about the environmental impact of conventional fossil fuel-based power plants have led to the desire to produce large amounts of renewable energy [7,23].

The most significant applications of modern technologies integrated into the AEDNs are: the advanced metering infrastructure, demand response, electric mobility, and distributed energy storage.

10.2.1   Advanced meter infrastructure

Using the smart technologies integrated into the advanced meter infrastructure (AMI) helps utility agencies and customers manage their consumption [24]. This infrastructure developed from an automatic meter reading (AMR) system improved with new functions (AMR plus), as seen in Figure 10.5.

A smart meter is an electronic device designed to communicate the values of a facility's energy consumption to its utility providers. It eliminates the need for manual meter readings and provides the facility with more granular details about its energy usage.

Regarding the meters included in an AMI, they can communicate with various head-end systems, such as those used by utilities. The ability of the AMI to provide real-time visibility represents a key component of AEDNs because it enables organizations to control their operations and improve efficiency.

SMART METER SYSTEM	FUNCTIONALITY	APPLICATIONS
Advanced metering infrastructure (AMI)	• Integrated service switch • Time-based rated • Remote meter programming • Power quality • Home area network (HAN) interface	• Marketing and demand side management • Load forecasting • Power procurement • Unregulated services
Automatic meter reading (AMR) PLUS	• Daily or on demand reads • Hourly interval data • Outage notification • Other commodity reads	• Distribution operations • Information technology • Metering services
Automatic meter reading (AMR)	• Automated monthly reads • One-way outage detection • Tamper detection • Load profiling	• Meter reading • Customer services & field services • Billing, accounting, collections

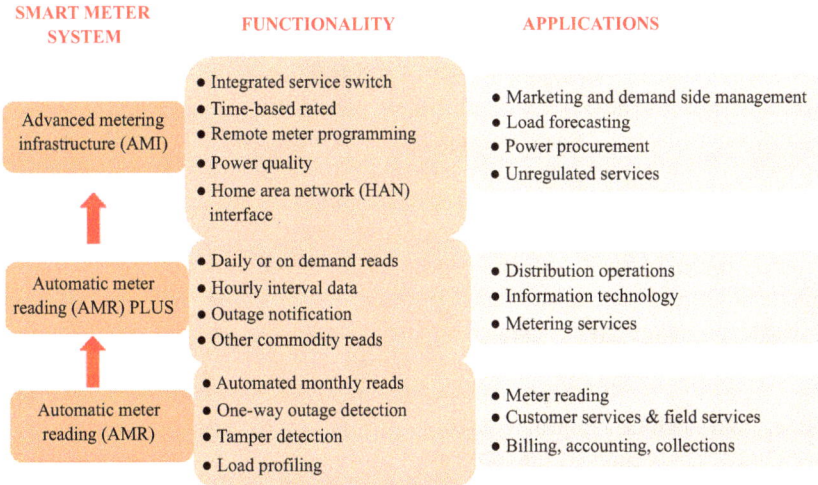

Figure 10.5   SM technology development according to [25]

More DNOs have stated that AMI represents the spine of the digital infrastructure that can accelerate the energy transition through their functionality and the implemented applications. The main objective for the roll-out of AMI is to improve the efficiency of the distribution infrastructure. It will allow the producers and the DNOs to optimize their operations and reduce the risk of over-investments. On the other hand, AMI can lead to indirect benefits for consumers. The third-generation smart meter facilitates dynamic pricing for SMEs and households, which allows them to reduce their energy bills [26]. Despite the various advantages of smart meters, most Member States do not consider addressing fuel poverty and energy efficiency as the main drivers for the roll-out of these technologies. For instance, in Romania, the risk of fuel poverty has recently reached alarming levels [7]. Some drawbacks can prevent consumers and DNOs from fully benefiting from these features. These include security issues and privacy concerns.

10.2.2   Distributed energy resources

This category of energy resource is usually renewable, generated near the point of use instead of being distributed through a centralized system. The wind and rooftop photovoltaic (PV) systems are examples of distributed energy resources (DER) installed at the level of the end-users from the AEDNs, also called prosumers. These systems include batteries and converters to store and inject energy. The main issues solved by the DER include load shifting, peak shaving, and voltage regulation. Furthermore, they can help balance the supply and electricity demand [27,28].

The prosumers are citizens who can partially meet their energy needs using various renewable energy sources such as biomass generators, cogeneration systems, wind turbines, and PV panels. They can be energy cooperatives, industrial enterprises, social institutions, or commercial enterprises.

The maximum generation capacity of a prosumer, which it can achieve through a small-size renewable energy project, varies from country to country and depends on their requirements. For example, a prosumer from Estonia can have a maximum generation capacity of 3.6 kW. The situation is different in Belgium, where a prosumer can achieve a connection approval for a maximum value of 10 kW. Spain or Romania has a higher generation capacity, touching the maximum of 100 kW. According to the study presented in [29], the energy communities in Europe have tremendous potential to produce renewable energy. Thus, half of the electricity consumers in the EU will cover renewable energy production until 2050. It would allow them to meet 45% of their energy needs.

The new energy rights granted to EU communities and citizens have as a foundation increasing the potential to transform the energy landscape of Europe. There are over two million people currently involved in over 7,700 energy communities, and the total renewable capacity of these communities is estimated to be over 6.3 GW [2]. Recently, the number of prosumers who have installed PV systems has increased significantly. They are mainly associated with the low prices of PV panels and the support schemes in their local countries.

The prosumers can decrease the cost of their electricity by injecting it into the network. In most cases, this cost is smaller than the prices offered by the energy market. Between 2008 and 2015, the cost of PV systems significantly decreased, such that more consumers installed the generation systems, becoming the prosumers. Incentives are also available to encourage the use of renewable energy sources. They can benefit from some advantages of having a PV system, such as lower electricity costs and increased energy independence [27,30]. Another factor that can influence the decision to become a prosumer is represented by the technological advancements in the industry, with a significant contribution to the increasing popularity of PV panels. Thus, the prosumers can represent a feasible solution to the global warming problem. The action plan aims to limit the temperature rise to below 2 °C compared with the value from 1990, meaning a significant step toward achieving sustainable development [29].

An end-user should go through several stages associated with the connection process at the network to obtain the status of the prosumer. The first stage is represented by the information and documentation process when an individual/community submits the request for a connection to the distribution network. This stage involves getting the necessary documents to obtain technical approval. The DNO establishes a connection solution to the generation system installed. Aside from the installation, the final steps include releasing the connection certificate and power-up the generation system [31].

One of the most important stages highlighted above contains the connection strategy developed at the DNO level, which should include all technical factors that affect the operation of the network (the length and cross-section of each branch, the allocation and connection phase of the 1-P end-users, including both consumers and prosumers already connected, at each pole), ensuring that the connection is successful and without to affect the AEDN strength.

10.2.3 Demand response

The demand response (DR) concept refers to program packages designed to help AEDNs to maintain their capacity. These programs are carried out through the voluntary actions of end-users and companies. Exploiting the potential of DR can help lower the power grid's load profile and help reduce greenhouse gas emissions. For example, about 800 TWh from processes with DR potential have been estimated for the EU's 28 member states at the level of 2012, representing almost 29% of the electricity consumption. Approximately 36% of the electricity generated by processes with DR potential corresponds to residential and tertiary sectors. According to estimations done in [32], the total potential of DR in Europe is 52.35 GW, of which approximately around 42% of the electricity generated by processes that have DR potential belongs to applications from the residential sector, 31% for the industrial sector, and 27% for tertiary sectors. Figure 10.6 presents the allocation for residential and tertiary sectors with a significant share in the AEDNs.

Such a program is a type of demand management that involves providing services to help improve the reliability of the AEDNs. Some of these include monitoring and controlling the power flows. Due to the development of policies and the establishment of open wholesale markets, DR programs can become more prevalent inside the smart technologies implemented in the AEDNs.

Two types of DR programs have been designed for customers. The price-based DR program [33] can manage the electricity usage by the consumers. It allows them to adjust their consumption to save money on their bills. One type of DR program is an incentive-based model [34]. This type of program encourages customers to adopt a more sustainable consumption pattern. Through the programs, customers can also help improve the reliability of the AEDNs by reducing their consumption during the peak load. This type of DR program is very beneficial for the power system.

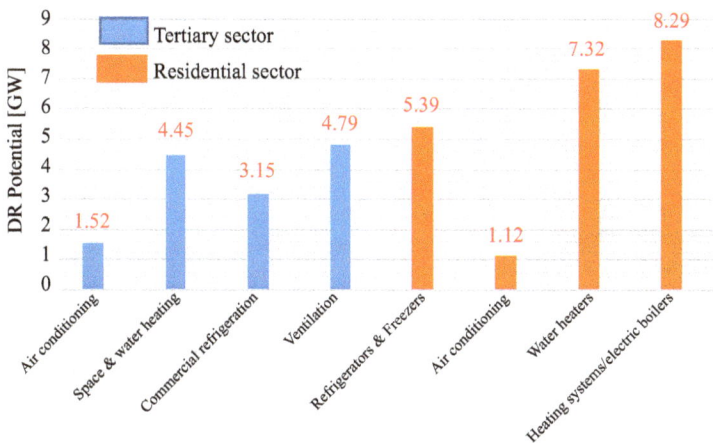

Figure 10.6   The DR potential for the residential and tertiary sectors from the UE according to [32]

10.2.4   *Electric mobility*

Despite the challenges associated with the electrification of vehicle parks, there are still many opportunities. The opportunities refer to the rapid emergence and evolution of mobility technologies, which created a dazzling array of new solutions for urban roads, including the development of mobility-as-a-service models, parking systems, and freight-sharing programs. Also, the availability of charging stations is a key aspect of the transportation infrastructure that will support the adoption of electric vehicles. Thus, a country's charging infrastructure should be designed to meet the needs of its citizens to their requests [35].

In Europe, the distribution of these facilities is not uniformly distributed, with the growing number of electric vehicle charging stations in the member countries. For instance, in Europe, 70% of all electric vehicle charging stations are located in Germany, France, and The Netherlands. According to [36], the number of electric vehicles charging stations will increase from approximately 7.9 billion in 2022 to over 32 billion in 2032. For example, on average, the EU offered five fast public chargers for every 100 km in 2021, see Figure 10.7 [37].

As the number of electric vehicles continues to increase, DNOs need to integrate into the AEDNs the latest innovative technologies to improve their operations and provide electric vehicle charging stations. However, issues regarding safety and congestion will appear in the big cities where more vehicles will add, and the population will increase [38]. Thus, the DNOs must prepare for the future of e-mobility by establishing the necessary infrastructure to support the growing demand for electric vehicles.

All these facts must be considered in the operation and planning strategies by the DNOs considering the influence on the AEDN strength. Regarding their

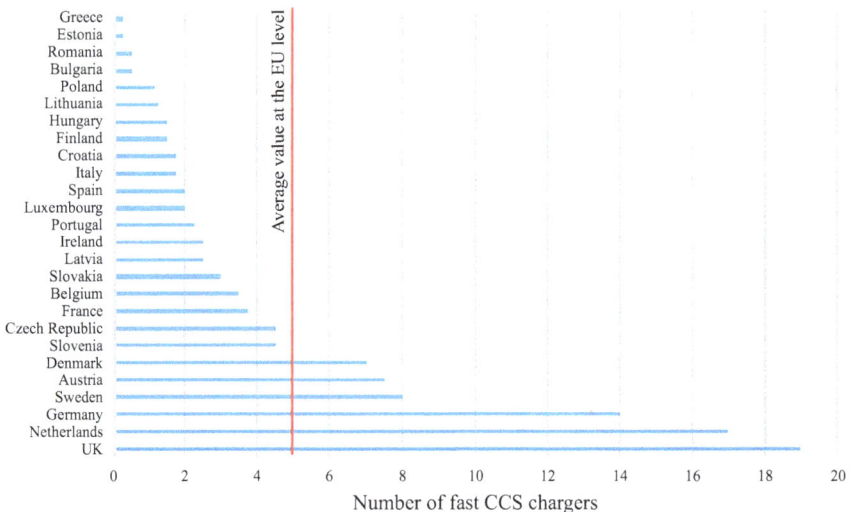

Figure 10.7   The number of fast chargers for every 100 km according to [37]

potential to sustain the AEDNs, the lack of flexibility in the load variables associated with these vehicles could threaten the AEDN strength. Fortunately, they can successfully manage the AEDNs using smart device-based control strategies. One of the main advantages of AEDNs is the ability to integrate the multiple storage resources of EVs.

10.3 Smart devices-based strategy in the optimal planning and operation of the AEDNs

Small-size renewable energy sources represented mainly by PV systems have become an EU-wide concern regardless of the country. The high penetration levels have opened the gate to the rapidly approaching future of active distribution networks. The gradual penetration of PV systems into the EDNs led to the concern of the DNOs demonstrated through multiple emerging technical issues (voltage, demand, protection, and power quality) [39]. These have forced the DNOs to initiate plans associated with prosumers' efficient integration strategies to minimize the impact and increase the hosting capacity of the EDNs such that the AEDN strength is improved. Many concerns and issues identified in the AEDNs with different PV penetration degrees have been addressed in the various studies where researchers proposed solutions, including strategies with discussion and highlighting their potential to be implemented. Although these strategies are not ideal, they improved the optimal operation and planning of the AEDNs. However, from the state-of-the-art performed by the authors regarding the proposals developed, none used together the smart meters (SMs), the phase load balancing (PLB) devices, and the OLTC to improve the system strength in the context of the large-scale integration of prosumers. The proposal of a smart device-based strategy and demonstration of its efficiency in keeping the AEDN strength represents the main target of this chapter.

The proposed strategy is based on the data provided by the SMs integrated at the level of the end-users (consumers and prosumers) [40] and the control of the PLB smart devices, installed to the prosumers having single phase (1-P) branching connected to one of the three phases [41], and the OLTC of the transformer from the EDS [42]. Through its implementation, the best solutions regarding the operation and planning of the AEDN are identified, considering the current situation (topology and the end-users connected at the phases) and various scenarios associated with the increasing trend in the number of prosumers quantified through a penetration degree calculated as the ratio between the number of prosumers and the total number of the end-users.

Figure 10.8 shows the flowchart of the proposed strategy containing four stages implemented inside a decision-making support system with two modules (database module and decision-making module). The details on each stage performed in each module are presented in the following.

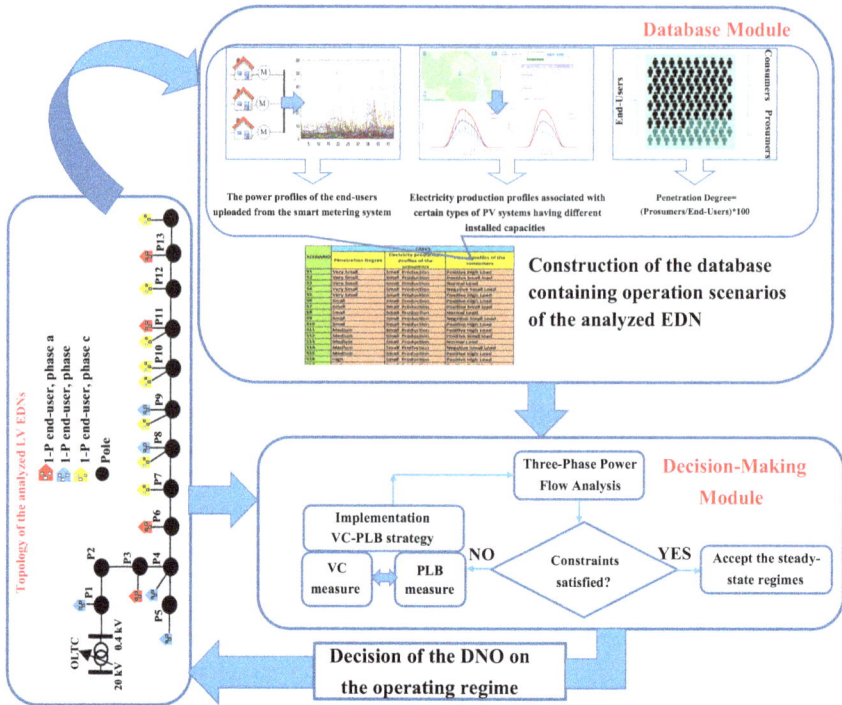

Figure 10.8   The flowchart of the VC-PLB strategy

10.3.1   Database module

This module manages two stages regarding the building of two databases: the topology database and injected/requested power profiles database.

Stage 1. Topology database

Technical characteristics of the lines (the cross-sections of the phase and neutral conductors, length, and the maximum allowable current) and the power transformer from the EDS (the rated power, if it is fitted or not with OLTC, the number of the tap positions) are uploaded from the network topology database. The longitudinal (resistance and reactance) and transversal (capacitive susceptance) characteristic parameters of each component (electric line or power transformer) identified between two poles (input, p_i, and output, p_o, with p_i, p_o $\in \{N\}$, $i \neq o$, where $\{N\}$ represents the set of the poles from the analyzed LV AEDN) are calculated considering the specific values indicated by the manufacturer.

Stage 2. Injected/requested power profiles' database

The database containing the injected/requested power profiles of the end-users (consumers and prosumers) uploaded from the smart metering system (SMS), electricity production profiles associated with certain types of PV systems having different installed capacities (uploaded from the PVGIS tool [43]), and the various feasible penetration degrees. The injected power profiles belong to the prosumers already connected to the AEDN, and the requested power profiles are associated with the consumer. Thus, the penetration degree can be initially 0 (all end-users are consumers) or the percentage of the prosumers from the total number of end-users.

10.3.2 Decision-making module

This module authorizes the decision-making process in the AEDN. The decision-making process regarding the operating regime is based on the steady-state calculation performed at the time slices t, $t = 1$, ..., T, where T is the analysis period and triggering the procedure for applying the VC-PLB strategy if the phase voltage constraint at the level of the poles p, $p = 1$, ..., N, in the AEDN is not satisfied. The process requires the following four stages: processing the input data, the steady-state calculation, verifying the voltage constraints, and the application procedure of the VC-PLB strategy.

Stage 3. Processing the input data for the steady-state calculation

The injected/requested powers of all end-users (consumers and prosumers) connected to each pole p, $p = 1, \ldots, N$, on each phase are aggregated. The injected power by the prosumer used the conventional minus sign ($-$), and the requested power by the consumers, the plus sign ($+$):

$$P_p^{\{\phi\}} = \sum_{e=1}^{N_{e,p}^{\{\phi\}}} P_{p,e}^{\{\phi\}} \quad p = 1, \ldots, N, \phi \in \{a, b, c, \, abc\} \tag{10.1}$$

$$Q_p^{\{\phi\}} = \sum_{e=1}^{N_{e,p}^{\{\phi\}}} Q_{p,e}^{\{\phi\}} \quad p = 1, \ldots, N, \phi = \{a, b, c, \, abc\} \tag{10.2}$$

where

$N_{e,p}^{\{\phi\}}$ represents the total number of end-users connected to the pole p, $p = 1$, ..., N, on the phases $\phi = \{a, b, c\}$, if it has a 1-P branching, and on all three phases $\phi = \{abc\}$ if it is a 3-P branching, from the analyzed AEDN.

For other scenarios associated with different penetration degrees, in the case of a new prosumer w connected at the pole p, the following relation is used for the injected/required power:

$$P_{p,w}^{\{\phi\}} = P_{p,w}^{c,\{\phi\}} - P_{p,w}^{IC,\{\phi\}} \quad p = 1, \ldots, N, \phi \in \{a, b, c, \, abc\} \tag{10.3}$$

$$Q_{p,w}^{\{\phi\}} = Q_{p,w}^{c,\{\phi\}} - Q_{p,w}^{IC,\{\phi\}} \ o = 1, \ \ldots, N, \phi \ \in \ \{a, b, c, \ abc\} \tag{10.4}$$

where $P_{p,w}{}^{\{\phi\}}$, $Q_{p,w}{}^{\{\phi\}}$ are the injected/requested powers (active and reactive) by the prosumer w connected at the pole p; $P_w{}^{c,\{\phi\}}$, $Q_w{}^{c,\{\phi\}}$, represents the self-consumption (active and reactive) of the prosumer w; $P_w{}^{IC,\{\phi\}}$ indicates the active power produced by the PV system installed at the prosumer w having the installed capacity IC. In practice, PV inverter operate with a value of power factor $\cos\phi = 1$, thus that the reactive power $Q_{p,w}{}^{IC,\{\phi\}}$ can be neglected.

In the case of each scenario associated with a certain penetration degree, it considers that those end-users who have the highest annual energy consumption (up to a certain threshold set by the decision maker (DM)), in descending order, will form the future prosumer set $\{N_{FP}\}$, thus $w\in\{N_{FP}\}$. Also, the same energy production profiles (uploaded from the PVGIS tool [43] in Stage 2) are used for the PV systems having similar installed capacities due to the end-users being very close and located in a geographical area with the same solar characteristics [44]. Not least, it is considered that new prosumers w, $w \in\{N_{FP}\}$, will use the same connection phase associated with the actual branching. This approach is very common to Romanian DNOs, which do not currently apply an optimization process for the optimal connection on the phases of the AEDNs of the prosumers.

Stage 4. The steady-state calculation

This calculation is based on the input data processed in Stage 2 and considers the determination of the following variables: phase voltages at the level of each pole p, $p = 1, \ldots, N$, the power flow and power losses in phase and neutral conductors of each line section between two poles (input, p_i, and output, p_o, with $p_i, p_o \in \{N\}$, $i \neq o$, where $\{N\}$), and the power on each phase $\{\phi\} \in \{a, b, c\}$ at the level of the LV side of the EDS (slack bus). An efficient forward/backward sweep-based algorithm, which has been proposed especially for the three-phase LV AEDN operating in the balanced and unbalanced regimes [40], is used due to their performance (accuracy and calculation time).

Stage 5. Verifying the voltage constraints

The constraint regarding the phase voltages associated with falling within the allowable limits specified in the performance standards is verified. If the constraint is violated even at the level of a single pole, a warning message will alert the decision-maker to initiate the procedure to trigger the application procedure of the voltage-control and phase load balancing (VC-PLB) strategy.

Stage 6. Application procedure of the VC-PLB strategy

The procedure contains two measures integrated inside the strategy and applied beginning with the VC measure and followed by the PLB measure (if this is required):

Measure 1. Voltage control is based on the OLTC with which the transformer from the EDS is fitted. An optimization process to determine the optimal tap position of the OLTC at each time slice t (usually 1 h) is initiated using a mathematical model having the following structure [42]:

- **Objective function**

$$\min\{\Psi(V,\omega)\} = \min\left\{\beta_1 \cdot \frac{\Delta P_{t,\omega}}{\Delta P_{t,\omega(0)}} + \beta_2 \cdot \frac{\Delta V_{t,\omega}}{\Delta V_{t,\omega(0)}}\right\},$$

$$\omega = 1,...,\Omega, \quad t = 1,...,T$$

(10.5)

$$\beta_1 + \beta_2 = 1$$

(10.6)

where

Ω is the number of the tap positions with which OLTC is fitted; ω is a certain tap position of the OLTC inside the range $[1, \Omega]$, where 1 is the minimum tap position and Ω is maximum tap position; $\omega(0)$ represents the constant tap position of the OLTC maintained constant for any time slice, t, $t = 1, \ldots, T$ without violating the phase voltage constraints; $\Delta P_{t,\omega}$ is the total power losses resulted from the steady-state calculation run for each time slice t, $t = 1, \ldots, T$, for a certain tap position ω of the OLTC, $\omega = 1,\ldots, \Omega$; $\Delta V_{t,\omega}$ is the sum of the phase voltage deviations calculated against the rated voltage for each time slice t, $t = 1, \ldots, T$, for a certain tap position ω of the OLTC, $\omega = 1,\ldots, \Omega$; β_1 and β_2 are the weighting coefficients associated with the importance degree of the two criteria integrated inside the objective function (total losses and the sum of voltage deviations).

- **Equality constraints**

  The power balance

  The relations (10.1) and (10.2) corresponding to the power balance are used inside the mathematical model.

- **Inequality constraints**

  Allowable limits for the phase voltage

  $$\underline{V} \leq V_{p,t,\omega}^{\{pf\}} \leq \overline{V}, \quad \phi \in \{a,b,c\}, \quad \omega = 1,...,\Omega,$$

  $$p = 1,...,N, \quad t = 1,...,T$$

  (10.7)

  where $\underline{V}$ and $\overline{V}$ are the allowable limits (lower and upper) of the phase voltage ($\pm 10\%$ from the rated voltage, $V_r = 230$ [45]); $V_{p,t,\omega}^{\{\phi\}}$ is the phase voltage determined as a result of steady-state calculations the level of the pole p, $p = 1, \ldots, N$, at the time slice $t = 1, \ldots, T$, for a certain tap position ω of the OLTC, $\omega = 1,\ldots, \Omega$.

  Thermal limits of line sections:

  $$S_{s,t,\omega}^{\{\phi\}} \leq S_s^{\max}, \quad \phi \in \{a,b,c\}, \quad s = 1,...,N_s,$$

  $$\omega = 1,...,\Omega, \quad t = 1,...,T$$

  (10.8)

where

S_s^{max} represents the maximum allowable apparent power of the line section s, $s = 1, \ldots, N_s$, of the AEDN; $S^{\{\phi\}}{}_{s,t,\omega}$ corresponds to the apparent power flow on each phase conductor of the line section s, $s = 1, \ldots, N$, determined as a result of steady-state calculations the level of the pole p, $p = 1, \ldots, N$, at the time slice $t = 1, \ldots, T$, for a certain tap position ω of the OLTC, $\omega = 1, \ldots, \Omega$.

Allowable limits of the tap position:

$$1 \leq \omega_t \leq \Omega, \quad t = 1, ..., T \tag{10.9}$$

where

ω_t is the tap position of the OLTC at the time slice t, $t = 1, \ldots, T$.

The optimal solutions of the mathematical model are obtained using the exhaustive search method proposed in [41]. Although it could lead to many computation iterations in some cases, the DM can consider only representative values of importance degrees associated with each criterion to reduce the search space. The optimal solution for the selected values of weighting coefficients, β_1 and β_2, is defined by the tap position, $\omega_t^{(opt)}$ of the OLTC at each time slice, t, $t = 1, \ldots, T$. It must satisfy the constraints (10.7)–(10.9) and minimize the objective function (10.5). If some constraints are not satisfied, then the second measure regarding the PLB process based on the smart device installed at the prosumers is activated, and after that, this measure will be repeated.

Measure 2. The PLB process is applied to swap the 1-P prosumers connected to the AEDN from one phase to another. The technical constraints related to the powers on each phase $\{\phi\} \in \{a, b, c\}$, such as the number of end-users and the location of each pole, p, $p = 1, \ldots, N$ and phase, are considered. The input data from Stage 2 are used to execute the PLB procedure for each time slice t, $t = 1, \ldots, T$. The procedure aims determination of the minimum value of the unbalanced factor, UF, at the level of each pole, p, $p = 1, \ldots, N$.

The components of the mathematical model are the following [41]:

- **Objective function**

$$\min(UF_{p,t}), \quad t = 1, ..., T, \quad p = 1, ..., N \tag{10.10}$$

$$UF_{p,t} = \frac{\displaystyle\sum_{\phi \in \{a,b,c\}} \left(\frac{P_{p,t}^{\{\phi\}}}{P_{av,p,t}}\right)^2}{n_\phi}, \quad t = 1, ..., T, \quad p = 1, ..., N \tag{10.11}$$

$$P_{av,p,t} = \frac{\displaystyle\sum_{\phi \in \{a,b,c\}} P_{p,t}^{\{\phi\}}}{n_\phi}, \quad t = 1, ..., T, \quad p = 1, ..., N \tag{10.12}$$

- **Equality constraints**

$$P_{p,t}^{\{\phi\}} = P_{C,p,t}^{\{\phi\}} + P_{P,p,t}^{(\phi)} + P_{(p-1),t}^{\{\phi\}}, \quad t = 1, \ldots, T, \quad p = 1, \ldots N, \quad \phi \in \{a, b, c\}$$

$$(10.13)$$

$$P_{C,p,t}^{\{\phi\}} = \sum_{l=1}^{N_{C,p,t}^{\{\phi\}}} P_{l,p,t}^{\{\phi\}}, \quad t = 1, \ldots, T, \quad p = 1, \ldots N, \quad \phi \in \{a, b, c\} \quad (10.14)$$

$$P_{P,p,t}^{\{\phi\}} = \sum_{w=1}^{N_{P,p,t}^{\{\phi\}}} a_{w,p,t}^{\{\phi\}} \cdot P_{w,p,t}^{\{\phi\}}, \quad t = 1, \ldots, T, \quad p = 1, \ldots N, \quad \phi \in \{a, b, c\}$$

$$(10.15)$$

$$N_{C,p} = \sum_{\phi \in \{a,b,c\}} N_{C,p,t}^{\{\phi\}}, \quad t = 1, \ldots, T, \quad p = 1, \ldots N \quad (10.16)$$

$$N_{P,p} = \sum_{\phi \in \{a,b,c\}} N_{P,p,t}^{\{\phi\}}, \quad t = 1, \ldots, T, \quad p = 1, \ldots N \quad (10.17)$$

$$N_{e,p} = N_{C,p} + N_{P,p}, \quad p = 1, \ldots N \quad (10.18)$$

$$N_e = \sum_{p=1}^{N} N_{e,p}, \quad p = 1, \ldots N \quad (10.19)$$

where

$N_{C,p}$ and $N_{P,p}$ are the number of the consumers and prosumers connected to the pole p of the AEDN, p, $p = 1, \ldots, N$.

$N_{e,p}$ represents the total number of the end-users (prosumers and consumers) connected to the pole p of the AEDN, p, $p = 1, \ldots, N$.

$P_{p,t}^{\{\phi\}}$ is the absorbed/injected active powers by the end-users $N_{e,p}$ connected on the phase $\{\phi\} \in \{a, b, c\}$, at the level of the pole p, $p = 1, \ldots, N$, and the time slice t, $t = 1, \ldots, T$.

$P_{C,p,t}^{\{\phi\}}$ is the absorbed active powers by the consumers l, $l = 1, \ldots, N_{C,p,t}^{\{\phi\}}$ connected on the phase $\{\phi\} \in \{a, b, c\}$, to the level of the pole p, $p = 1, \ldots, N$, and the time slice t, $t = 1, \ldots, T$.

$P_{P,p,t}^{\{\phi\}}$ is the absorbed/injected active powers by the prosumers w, $w = 1, \ldots, N_{P,p,t}^{\{\phi\}}$ connected on the phase $\{\phi\} \in \{a, b, c\}$, to the level of the pole p, $p = 1, \ldots, N$, and the time slice t, $t = 1, \ldots, T$.

$P_{(p-1),t}^{\{\phi\}}$ the active powers on the phases ϕ, $\{\phi\} \in \{a, b, c\}$, aggregated at the pole $(p + 1)$ downstream by the pole p, $p = 1, \ldots, N$, and the time slice t, $t = 1, \ldots, T$. In the first iteration, when the start the PLB algorithm, the values are equal with 0 for the end poles.

$a_{\{p\},l}^{(t)}$ represents a binary variable (0 or 1) defined to determine if a prosumer w, $w = 1, \ldots, N_{P,p,t}^{\{\phi\}}$ is connected or not on the phase ϕ at the pole p, $p = 1, \ldots, N$, and time slice t, $t = 1, \ldots, T$.

The algorithm will start in the first iteration with the end poles of each lateral branch and main trunk and will stop with the low-voltage bus of the EDS in the last iteration. At the level of each pole p, $p = 1, \ldots, N$, the optimal solution represented by the connection phase of each prosumer is determined using combinatorial optimization [46].

After that, Measure 1 is repeated to identify the optimal tap position of the OLTC at each time slice t (usually, 1 h).

Stage 4. Communicate the optimal solution to the OLTC and PLB devices

Communicate to the PLB smart devices installed to the prosumers new connection phases (if it is necessary) and the OLTC the optimal tap position.

10.4   Testing the strategy

A three-phase low-voltage AEDN with 88 poles supplying a rural area belonging to a Romanian DNO was used to test and demonstrate the effectiveness of the proposed strategy. Figure 10.9 shows the topology of the test network, and Table 10.1 summarizes the characteristic technical data.

The AEDN is connected at an EDS equipped with an OLTC-fitted transformer having a rated power of 160 kVA. The OLTC has 9 tap positions, where the median tap is 5, with an adjustment voltage step of 2.5% between $[-10\%, +10\%]$ of the rated voltage of 230 V as specified in the European standard EN 50160 [45]. According to technical data presented in [39], the maximum number of operations guaranteed by the manufacturer is 700,000. The main trunk of the AEDN (MT_AEDN), between EDS and P88, has 53 poles with a single lateral branch (LB_AEDN) connected to pole P4 and having P39 as the end pole. Also, other branches with shorter lengths can be identified starting from the lateral branch P4–P39 with poles P8 and P23 as connecting nodes. The distance measured between the two poles is approximately 0.04 km, imposed through the design standard.

Table 10.2 presents the repartition of the end-users (consumers and prosumers) on the branch (MT_AEDN and LB_AEDN) and phase. The values expressed in percentages have been obtained by referring to the total number of end-users from the AEDN. It can be highlighted that the number of consumers allocated to the two branches is similar (51 compared to 52), and the number of prosumers is higher on the MT_AEDN branch. Phase b of the MT_AEDN branch has allocated more end-users connected than the LB_AEDN branch. Instead, the LB_AEDN branch has a higher number on phase c than the MT_AEDN branch. Also, a single three-phase (3-P) consumer is connected to the phases of the LB_AEDN branch. This unequal phase allocation of the end-users leads to a phase load unbalance with influences on the power losses and the voltage level.

Regarding the prosumers, they have installed the PV systems with the installed capacities of 3 and 5 kWp, without the energy storage batteries (ESBs). This prosumers type is characteristic of Romania, where very few prosumers also have the

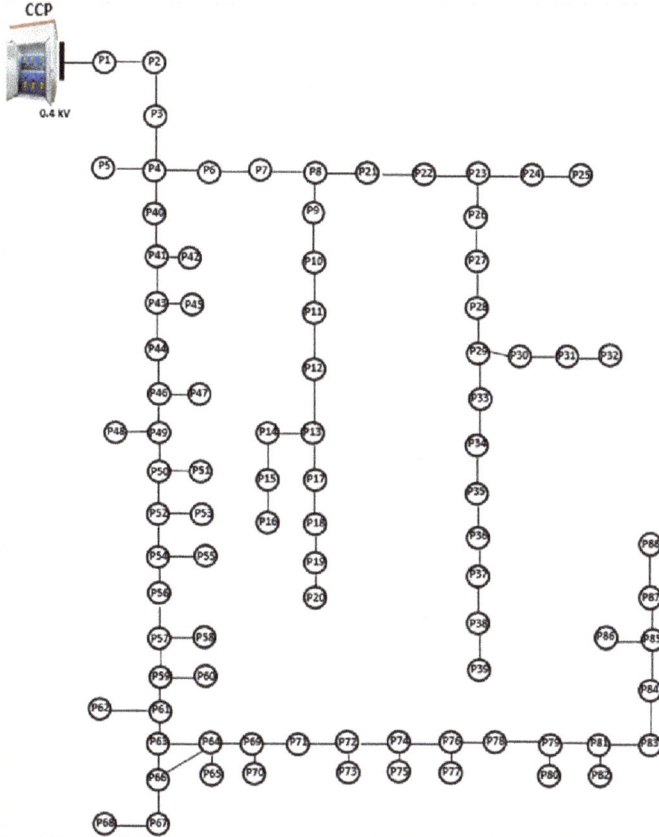

Figure 10.9   The topology of the test AEDN

Table 10.1   The characteristic technical data of the test AEDN

Branches	Number of poles	Initial pole	Final pole	Length (km)	Cross-section of the conductors (mm²)	
					Phase	Neutral
MT_AEDN EDS – P88	53	EDS	P88	2.04	3 × 50	50
		P63	P68	0.12	3 × 16	25
LB_AEDNP4 – P39	35	P4	P13	0.32	3 × 35	35
		P13	P20	0.28	1 × 35	35
		P8	P32	0.48	3 × 35	35
		P29	P39	0.28	1 × 25	25

Table 10.2   The repartition of the end-users on the branch and phase

Branches	Number of end-users	Number of consumers	Number of prosumers	Repartition of the phases			
				a (1-P)	b (1-P)	c (1-P)	abc (3-P)
MT_AEDN	61	52	9	8	51	2	0
EDS – P88	53.5%	45.6%	7.9%	7%	44.7%	1.8%	0%
LB_AEDN	53	51	2	8	4	40	1
P4 – P39	46.5%	44.7%	1.8%	7%	3.5%	35.1%	0.9%
Total AEDN	114	103	11	16	55	42	1
	100%	90.4%	9.6%	14.1%	48.2%	36.8%	0.9%

ESBs available due to the very high costs reported to the standard of living of the citizens and the lack of a coherent policy from the regulatory authorities. The DNO operates the AEDN with a fixed tap position (7 from 9) of the OLTC for all time slices, t, $t = 1, \ldots, 24$, and a 10% PV penetration degree (represented by the share of the 11 prosumers from the total number of the end-users) approximately.

The study has been performed considering the scenarios associated with various degrees of penetration of the prosumers, between 10% and 50%, representing shares from the total end-users. One of the assumptions used in the case of each penetration degree took into account that those end users who have the highest annual energy consumption, in descending order, will become prosumers. The surplus-generated power profiles are known for the prosumers integrated into the AEDN (11 prosumers). For the other end-users who can become prosumers in the analyzed scenarios, the energy production profiles of the PV systems with installed capacities of 3 and 5 kWp resulting from the PVGIS platform have been used from the day with maximum generated powers. This assumption has been considered because the end-users are very close and located in a geographical area with the same solar characteristics [35].

Figure 10.10 presents the generation profiles for the PV systems uploaded from the PVGIS platform [34] and assigned to the new prosumers. Also, another assumption considers that the new prosumers will use the same connection phase. This approach is similar to Romanian DNOs, which do not currently apply an optimization process for the optimal connection on the phases of the AEDNs of the prosumers.

Figure 10.11 shows the active power flows in the phase conductors on the EDS-P1 section of the MT_AEDS in the actual situation (10% degree penetration) considered as the base scenario, S0, and the simulated scenarios where the considered PV penetration degrees are 20% (23 prosumers), 30% (33 prosumers), 40% (46 prosumers), and 50% (57 prosumers) adopting the assumption of a constant tap position of the OLTC ($t = 7$) in the analyzed period ($T = 24$ h, with a time slice by 1 h). The analysis of the results highlighted that beginning with the 20% penetration degree, an inverse power active flow through the transformer from the EDS on all three phases. Also, an unbalance at the level of the power flows can be observed

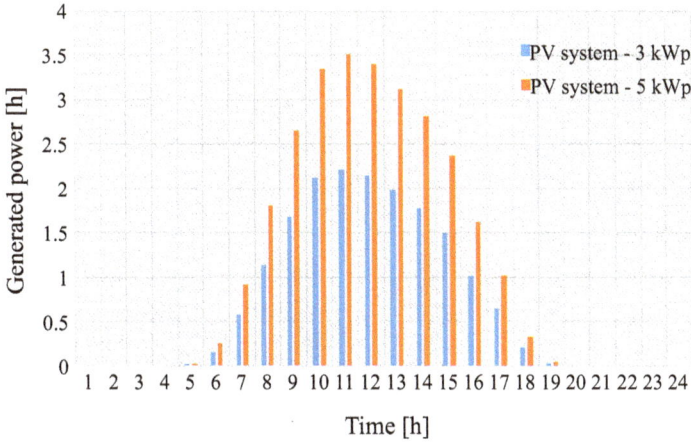

Figure 10.10 The generation profiles for the PV systems used in the analyzed scenarios

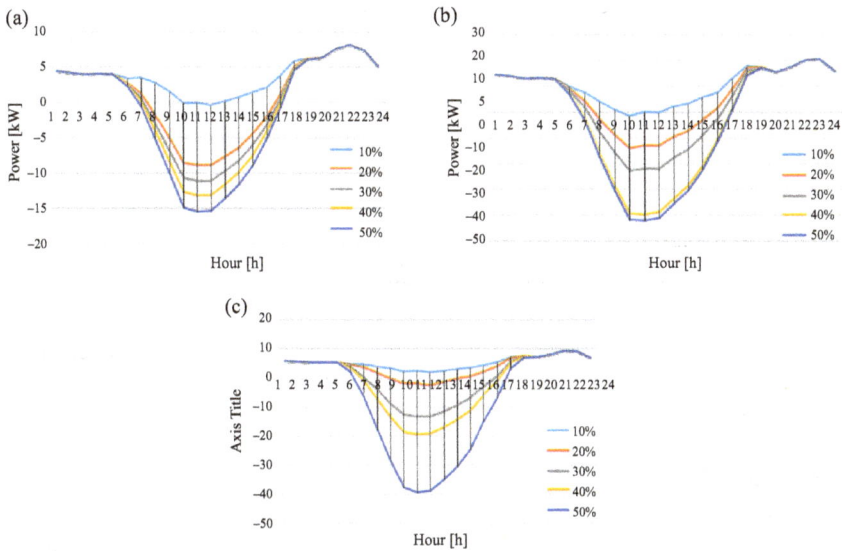

Figure 10.11 The active power flows in the phase conductors on the EDS-P1 section of the MT_AEDS – without application of the VC-PLB strategy (a – phase a; b – phase b; c – phase c)

that can be quantified through an average unbalance factor calculated for the analyzed period using the relation (10.11) and presented in Figure 10.12.

The calculated values of UF_{av} confirm the power unbalance, which is over 20% (between 22.9% and 31.5%) affecting the energy losses and the voltage level, see Table 10.3.

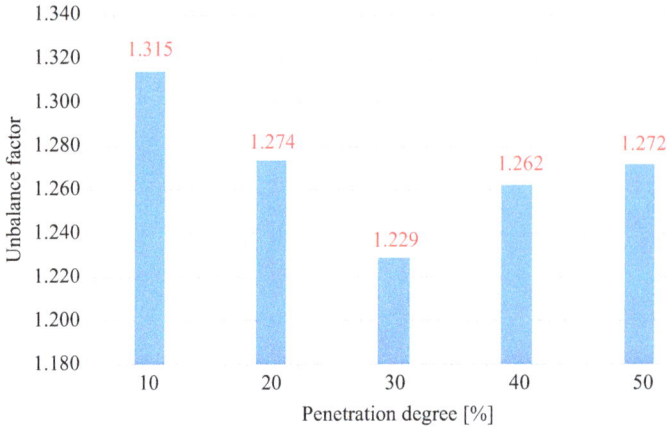

Figure 10.12   The average unbalance factor in the analyzed scenarios, without the application of the VC-PLB strategy

Table 10.3   The technical indicators regarding the energy losses and phase voltage obtained in the analyzed scenarios, without the application of the VC-PLB strategy

Variable	Scenarios associated with the penetration degree				
	S0 – 10%	S1 – 20%	S2 – 30%	S3 – 40%	S4 – 50%
Energy losses (kWh)	50.02	56.23	75.83	110.18	158.25
Maximum phase voltage (p.u.)	1.05	1.12	1.19	1.25	1.28

The data have been obtained based on the steady-state calculations performed with an efficient forward/backward sweep-based algorithm, especially for the three-phase LV AEDN operating in balanced and unbalanced regimes [31]. Regarding the energy losses, the values increase compared to the base scenario (S0 – 10%) with 12.4% for S1 – 20%, 51.6% for S2 – 30%, 120.3% for S3 – 40%, and 216.4% for S4 – 50%.

The maximum phase voltage increase over the allowable value accepted of 10%, beginning with the penetration degree of 20% when it exceeds 2% this value and finishing with the last considered penetration degree of 50%, where the overcome is very high, reaching 18%.

Table 10.4 presents the hours when the phase voltage is over on the maximum allowable value at the level of the poles from the AEDN. A first observation refers to the voltage constraints regarding the minimum allowable value, which are not violated. The hours when the phase voltage exceeds the maximum allowable corresponds to the time interval when the prosumers inject the power in the AEDN. A single hour from scenario S1 – 20% has been recorded when the voltage on phase b

Table 10.4 *The areas from the AEDN represented by the number of the poles affected by the exceeding the maximum allowable voltage on the phases*

Hour	Scenarios associated with the penetration degree											
	S1 – 20%			S2 – 30%			S3 – 40%			S4 – 50%		
	a	b	c	a	b	c	a	b	c	a	b	c
8	0	0	0	0	0	0	0	18	0	0	25	17
9	0	0	0	0	30	0	0	43	5	0	44	32
10	0	0	0	0	41	3	0	48	14	0	54	33
11	0	33	0	0	46	14	28	85	32	32	85	85
12	0	0	0	0	39	5	0	48	14	0	54	33
13	0	0	0	0	41	12	0	63	20	0	82	42
14	0	0	0	0	37	5	0	51	17	0	58	34
15	0	0	0	0	28	0	0	43	13	0	44	33
16	0	0	0	0	0	0	0	0	0	0	0	13

at the level of the 33 poles (37.5% of the total number of the poles) exceeds the maximum value.

The operating regime worsens in scenario S2 – 30% where the number of hours increases, being affected the phases b and c, with a higher number of poles at the level of phase b. For the last two scenarios, S3 – 40% and S4 – 50%, phase a of the AEDN will be affected, in addition to phases b and c, and the number of poles is very high (e.g., phase b presents the issues at the hour 11 in a percentage of 96.6 from the whole AEDN, 85 poles respectively).

Thus, the VC-PLB strategy associated with OLTC-based voltage control and PLB smart devices installed at the prosumers has been applied.

Figures 10.13, 10.14, and Table 10.5 present the results detailed above.

The analysis of Figure 10.13 highlighted that the inverse power active flows through the transformer from the EDS on all three phases are preserved. On the other hand, the power flows on the three phases in all scenarios are approximately equal due to the PLB devices installed at the prosumers leading to a reduction of energy losses. Also, the technical benefits refer to the average unbalance factor, UF_{av}, which has values below 10% (threshold accepted by the Romanian DNOs in the AEDNs), see Figure 10.14. Scenario S4 – 50% recorded the lowest value due to the higher number of combinations of the PLB devices available to determine the minimum value of the unbalanced factor.

Regarding the energy losses, these have half the values recorded without the implication of the VC-PLB strategy, see Table 10.5. Thus, energy saving has been between 49% and 54%. The lowest value has been recorded for scenario S4 – 50% at 49.01%, and the highest for scenario S2 – 30% at 53.94%.

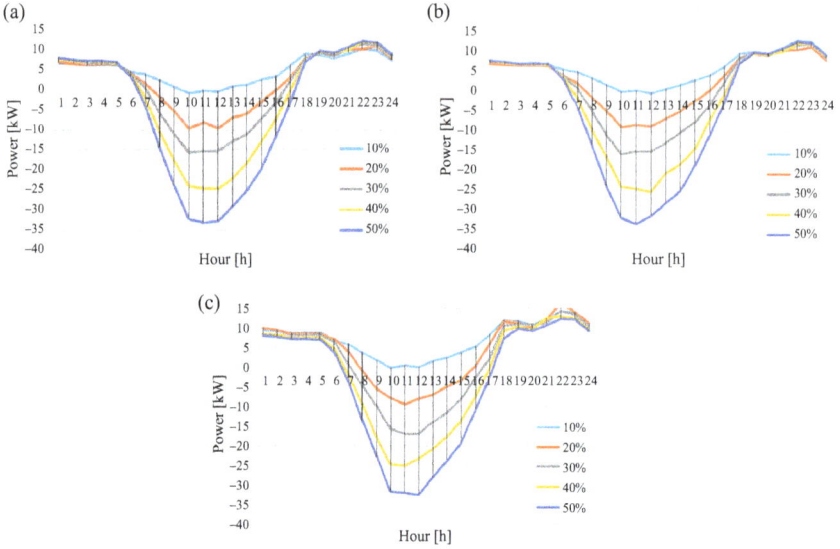

Figure 10.13   The active power flows in the phase conductors on the EDS-P1 section of the MT_AEDS – with the application of the VC-PLB strategy (a – phase a; b – phase b; c – phase c)

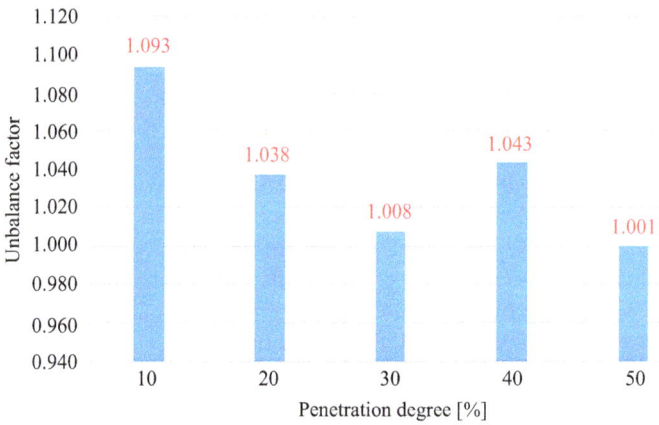

Figure 10.14   The average unbalance factor in the analyzed scenarios, with the application of the VC-PLB strategy

The phase voltages at the level of each pole from the AEDN have values below the maximum allowable limit due to the OLTC, which allowed the fulfilment of the phase voltage constraints, see Table 10.5.

Figure 10.15 presents the optimal tap positions of the OLTC at each hour, which led to an optimal voltage level in the AEDN. The minimum tap positions

Table 10.5 *The technical indicators regarding the energy losses and phase voltage obtained in the analyzed scenarios, with the application of the VC-PLB strategy*

Variable	Scenarios associated with the penetration degree				
	S0 – 10%	S1 – 20%	S2 – 30%	S3 – 40%	S4 – 50%
Energy losses (kWh)	24.58	26.23	34.92	53.74	80.68
Maximum phase voltage (p.u.)	1.05	1.05	1.05	1.06	1.06

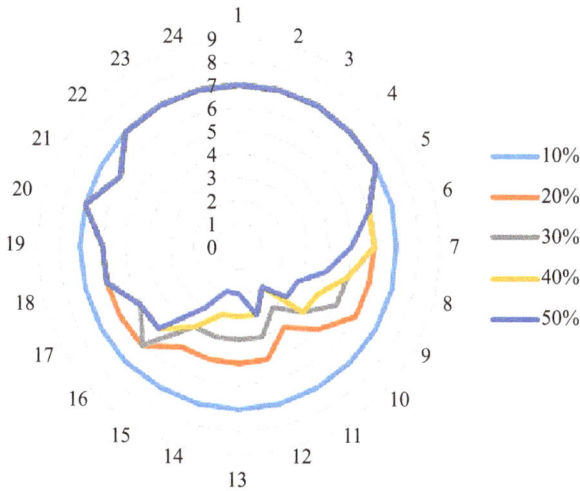

Figure 10.15 *The optimal tap positions of the OLTC with the implementation of the VC-PLB strategy for all analyzed scenarios*

have been recorded at hours 11 (scenarios S3 – 40% and S4 – 50%), 13 (scenario S4 – 50%), and 14 (scenario S4 – 50%) when the injected power by the prosumers had the maximum value.

The voltage quality has been analyzed based on the mean phase voltage unbalance rate (MPVUR) calculated at the level of each pole from the AEDN in the period $T = 24$ h, for which the following relation has been used according to the IEEE specifications [47,48]:

$$MPVUR_p \ [\%] = \frac{1}{T} \cdot \sum_{t=1}^{T} \left(\max_{\phi \in \{a,b,c\}} \left\{ \frac{\left| V_{p,t}^{\{\phi\}} - V_{av,p} \right|}{V_{av,p}} \cdot 100 \right\} \right), \qquad (10.20)$$

$$p = 1, \ldots, N$$

where $V_{av,p}$ represents the average value of the phase voltage at the level of the pole p, $p = 1, \ldots, N$.

Figure 10.16 presents the value of the MPVUR for each pole in the scenarios considered.

The values of MPVUR have been significantly reduced on the two branches MT-AEDS (EDS-P88) and LB-AEDS (P4-P39) if the VC-PLB strategy is applied, with a significant reduction at the end poles, which means a very high voltage quality. A decrease from 7.05% (scenario S0 – 10% without the VC-PLB strategy) to 2.43% (scenario S0 – 10% with the VC-PLB strategy) and 0.1% (scenario S2 – 30% with the VC-PLB strategy), has been recorded in the case of end pole P88 (MT-AEDS branch). The value of MVPUR at the end pole P39 of the LB-AEDS branch reduced from 2.77% (scenario S0 – 10% with applying the VC-PLB strategy) to 0.23% (scenario S4 – 50% with the VC-PLB strategy).

The efficiency of the proposed VC-PLB strategy can be highlighted through a comparison with the obtained results with the OLTC-based VC strategy proposed in [49], as seen in Table 10.6. The OLTC-based VC strategy can be efficient in terms of the voltage quality at the level of the poles from the AEDN until degree penetration of 30% (scenario S2 – 30%), for which the maximum value of the phase voltage has reached the upper allowable limit of 10%. Also, an improvement of the voltage quality can be observed for the strategies S3 – 40% and S4 – 50%, from 1.25 and 1.28 p.u. at 1.13 p.u., but the maximum phase voltage is over the upper allowable limit with 0.03 p.u.

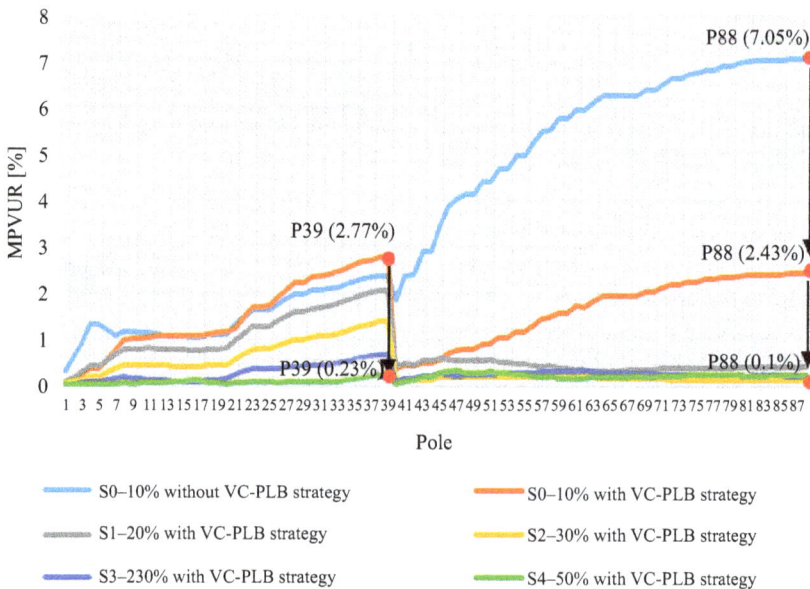

Figure 10.16   MPVUR calculated at the level of each pole for scenarios considered in the analysis

Table 10.6 The comparison between the phase voltages obtained in the analyzed scenarios with OLTC-based VC strategy and VC-PLB-based strategy

Approaches	Scenarios associated with the penetration degree				
	S0 – 10%	S1 – 20%	S2 – 30%	S3 – 40%	S4 – 50%
Without a strategy	1.05	1.12	1.19	1.25	1.28
With OLTC-based VC strategy	1.05	1.08	1.10	1.13	1.13
With VC-PLB-based strategy	1.05	1.05	1.05	1.06	1.06

10.5 Research gaps, challenges, and future research directions

In recent years, the electricity distribution infrastructure began to be the subject of accelerated development, putting forth sustained efforts to ensure the transition towards the AEDNs. The reduction in the energy production of centralized power plants due to the availability of renewable energy sources represented the main factor in the changes in the electricity distribution sector.

Thus, the classical EDNs are the subject of a radical modernization that allows this transition from passive to active features. The intermittent character of renewable energy, the high number of small-size power units located near consumers or even to consumers (prosumers) connected to the phases of the AEDNs, and the electro-mobility represented by electric vehicles can be associated with the main challenges for the DNOs.

The AEDNs concept represents the target of the modernization process of the classical EDNs that DNOs must be reached, integrating all categories and types of small-size power units into their operations based on innovative technologies. The growing number of PV systems installed at the end-users will continue to increase annually, according to the last reports of the energy regulatory agencies. Besides being more energy-efficient, these systems help in reducing greenhouse gas emissions.

Unfortunately, the lack of proper technical infrastructure and the increasing number of PV prosumer groups are not yet fully resolved due to, on the one hand, the varying topologies and the high percentages of the ageing components (transformers, lines, switching equipment, etc.) ensure the distribution infrastructure still represents the weaknesses of EDNs and, on the other hand, the most common factor that can affect the distribution capacity to accommodate the increasing number of solar photovoltaic systems represented by the voltage quality.

Thus, the high intermittency degree of these sources could be an issue for the DNOs in the voltage control from the LV EDNs as long as they do not use strategies to mitigate the voltage violations when a high number of PV prosumers are merged, in the situation in which the voltage control range is [+10%, −10%] of the rated voltage. Due to the increasing number of PV systems and the complexity and size of the LV EDNs, developing new operation and planning strategies must include more advanced technologies integrated into modern equipment, such as automatic voltage regulators, load balancing devices, energy storage systems, or capacitor banks, the newest data processing technologies based on smart devices that can quickly identify the optimal solutions and high-speed communication networks which to allow transmission of high data amounts and coordination in real-time all devices/equipment integrated into the network. In the face of these challenges, Artificial Intelligence techniques can represent the brain of the decision-making process to identify the solutions which lead to the best performance of the AEDNs. Thus, the solutions resulting from a decision-making process based on efficient and resilient operation and planning strategies can help the DNOs to improve the AEDN strength in the presence of the prosumers.

The digitalization process, representing the revolution of smart technologies, describes the future of the electricity distribution sector, characterized by resilience, efficiency, and reliability to ensure AEDN strength.

10.6   Conclusions

The gradual penetration of small-size RESs, represented mainly by PV systems, into the AEDNs led to the concern of the DNOs demonstrated through multiple emerging technical issues (voltage, demand, protection, and power quality) identified through distribution circuit measurements and analyses. In addition to the installation area, the connection point of the RESs can have a significant influence on the technical and economic performances of the AEDNs. These have forced the DNOs to initiate plans associated with efficient strategies to integrate the prosumers, minimizing the impact and increasing the hosting capacity of the AEDNs.

Thus, a new smart devices-based VC-PLB strategy for optimal operation and planning of the AEDNs integrating the small-size local renewable generation sources with various penetration degrees to improve the system strength has been proposed, demonstrating their practical implementation. Testing has been done in a Romanian AEDN considering the scenarios associated with various degrees of penetration of the prosumers, between 10% and 50% (S0 – 10%, S1 – 20%, S2 – 30%, S3 – 40%, and S4 – 50%) representing shares from the total end-users.

The analysis of the results highlighted that beginning with the 20% penetration degree, an inverse power active flow through the transformer from the EDS on all three phases and a high power unbalance have been identified. Regarding the energy losses, the values had a strong growth trend compared to the base scenario (S0 – 10%). The maximum phase voltage increased over the allowable value accepted of 10%, beginning with the penetration degree of 20%.

The application of the proposed strategy associated with OLTC-based voltage control and PLB smart devices installed at the prosumers led to the values approximately equal to the active powers on the three phases in all scenarios (quantified through values very close to the ideal target, 1.00 p.u., of UFav) due to the PLB devices installed at the prosumers with a positive impact on the reduction of energy losses, obtaining energy savings between 49% and 54% in the analyzed scenarios. The phase voltages at the level of all poles fell between the allowable limits due to the OLTC. Also, the improvement of the voltage quality quantified through the mean phase voltage unbalance rate (MPVUR) has been observed.

However, their implementation depends on the speed of data acquisition, communication network, and processing system for real-time data analysis and control.

The future work will consider an extension of the proposed strategy that considers the other control configurations, including the battery storage systems, demand response programs at the customers, and electric vehicle charging stations. The purpose is to include as many advanced technologies as possible with a high penetration degree into the future AEDNs.

References

[1]   European Environment Agency, Energy Prosumers in Europe Citizen Participation in the Energy Transition, 2022. https://www.eea.europa.eu/publications/the-role-of-prosumers-of.

[2]   European Commission, Report from the Commission to the European Parliament, the Council, the European Economic and Social Committee and the Committee of the Regions. State of the Energy Union 2021 – Contributing to the European Green Deal and the Union's Recovery, 2021. https://eur-lex.europa.eu.

[3]   D'adamo C., Abbey C., Jupe S., Buchholz B., Khattabi M., and Pilo F. 'Development and Operation of Active Distribution Networks: Results of Cigre C6.11 Working Group', *Proceedings of the 21st International Conference on Electricity Distribution*; Frankfurt, Germany, June 2011.

[4]   Neagu B., Grigoras G., and Ivanov O. 'The optimal operation of active distribution networks based on smart metering'. In Ustun T.S. (ed.), *Advanced Communication and Control Methods for Future Smartgrid*. London: IntechOpen; 2019.

[5]   Blasi T.M., de Aquino C.C.C.B., Pinto R.S., *et al.* 'Active distribution networks with microgrid and distributed energy resources optimization using hierarchical model'. *Energies.* 2022;15:3992.

[6]   Union of the Electricity Industry – Eurelectric, Distribution Grids in Europe Facts and Figures 2020, 2020. https://cdn.eurelectric.org/media/5089/dso-facts-and-figures-11122020-compressed-2020-030-0721-01-e-h-6BF237D8.pdf.

[7]   European Commission, Directorate-General for Energy, 'Benchmarking Smart Metering Deployment in the EU-28: Final Report'. Publications Office, 2020. https://data.europa.eu/doi/10.2833/492070.

[8]   European Commission, Smart Grids and Meters, 2022. https://energy.ec.europa.eu/topics/markets-and-consumers/smart-grids-and-meters_en#deployment-of-smart-meters.

[9]   Abrahamsen F.E., Ai Y., and Cheffena M., 'Communication technologies for smart grid: a comprehensive survey'. *Sensors.* 2021;21:8087.

[10] Dong J., Xue Y., Kuruganti, T., *et al.*, 'Operational impacts of high pene-
      tration solar power on a real-world distribution feeder'. *Proceedings of the
      IEEE Power & Energy Society Innovative Smart Grid Technologies
      Conference*, Washington, DC, July 2018. pp. 1–5.

[11] Wasiak I., Szypowski M., Kelm P., *et al.*, 'Innovative energy management
      system for low-voltage networks with distributed generation based on pro-
      sumers' active participation'. *Applied Energy*. 2022;312:118705.

[12] Urdal H., Ierna R., Zhu J., Ivanov C., Dahresobh A., and Rostom, D.,
      'System strength considerations in a converter dominated power system'.
      IET Renewable Power Generation. 2015;9:10–17.

[13] Gavrilovic, A. 'AC/DC system strength as indicated by short circuit ratios'.
      In *Proceedings of the International Conference on AC and DC Power
      Transmission*, London, UK, Sep 1991. pp. 27–32.

[14] ENTSO-E, 'NC HVDC – Call for Stakeholder Input', https://www.entsoe.eu/
      fileadmin/user_upload/_library/resources/HVDC/130507-NC_HVDC_-_Call_
      for_Stakeholder_Input.pdf.

[15] Xu Z., Zhang N., Zhang Z., and Huang Y. 'The definition of power grid
      strength and its calculation methods for power systems with high proportion
      nonsynchronous machine sources'. *Energies*. 2023;16:1842.

[16] AEMO, System Strength Impact Assessment Guidelines; Sydney, Australia,
      2020. p. 43.

[17] Yu L., Sun H., Xu S., Zhao B., and Zhang J. 'A critical system strength
      evaluation of a power system with high penetration of renewable energy
      generations'. *CSEE Journal of Power and Energy Systems*. 2022;8:710–720.

[18] Badrzadeh B. and Emin Z., 'System Strength'. https://www.cigre.org/article/
      GB/publications/reference-papers/system-strength.

[19] Dozein M.G., Mancarella P., Saha T.K., and Yan R. 'System strength and
      weak grids: fundamentals, challenges, and mitigation strategies'. In
      *Proceedings of the Australasian Universities Power Engineering Conference
      (AUPEC)*, Auckland, New Zealand, 2018. pp. 1–7.

[20] Yu L., Meng K., Zhang W., and Zhang Y. 'An overview of system strength
      challenges in Australia's national electricity market grid'. *Electronics*.
      2022;11:224.

[21] Shafik M.B., Rashed G.I., and Chen H., 'Optimizing energy savings and
      operation of active distribution networks utilizing hybrid energy resources
      and soft open points: case study in Sohag, Egypt'. *IEEE Access*.
      2020;8:28704–28717.

[22] Mondejar M., Avtar R., Baños Diaz H.L., *et al.* 'Digitalization to achieve
      sustainable development goals: steps towards a Smart Green Planet'. *Science
      of the Total Environment*. 2021;794:148539.

[23] Oladeji I., Makolo P., Abdillah M., Shi J., and Zamora R. 'Security impacts
      assessment of active distribution network on the modern grid operation—a
      review'. *Electronics*. 2021;10(16):2040.

[24]   Edison Electric Institute. 'Smart Meters and Smart Meter Systems: A Metering Industry Perspective', 2011. https://aeic.org/wpcontent/uploads/2013/07/smartmetersfinal032511.pdf.

[25]   Burchill A. Smart Metering and Its Use for Distribution Network Control. Cardiff University: Doctor of Philosophy Thesis, 2018.

[26]   Shafique H., Bertling Tjernberg L., Archer D.E., and Wingstedt S. 'Behind the meter strategies: energy management system with a Swedish case study'. *IEEE Electrification Magazine.* 2021;9(3):112–119.

[27]   Centre for Sustainable Energy Studies, Prosumers' Role in the Future Energy System, 2018. https://www.ntnu.edu/documents/1276062818/1283878281/.

[28]   Sirviö K., Berg P., Kauhaniemi K., Laaksonen H., Laaksonen P., and Rajala A., 'Socio-technical modelling of customer roles in developing low voltage distribution networks'. In *Proceedings of the Cired Workshop*, Ljubljana, Slovenia, June 2018.

[29]   Ballesteros M. 'The European prosumer: the unknown creature'. In *7th Vienna Forum on European Energy Law*, Vienna, Austria, Sept 2019.

[30]   GreenMatch, Can Smart Meters Work with Solar Panels? How Smart Meters Help Monitor Your Energy Consumption, 2021. https://www.greenmatch.co.uk/blog/smart-meters-and-solar-panels.

[31]   Romanian Energy Regulatory Authority, Order 19/2022 for the Approval of the Procedure Regarding the Connection to the Electricity Networks of Public Interest of the Places of Consumption and Production Belonging to the Prosumers (in Romanian), 2022. www.anre.ro.

[32]   SiaPartners, Demand Response: A Study of its Potential in Europe, 2014. https://www.sia-partners.com/en/news-aFnd-publications/from-our-experts/demand-response-a-study-its-potential-europe.

[33]   Judge M.A, Manzoor A., Maple C., Rodrigues J., and Islam S. 'Price-based demand response for household load management with interval uncertainty'. *Energy Reports.* 2021;7:8493–8504.

[34]   Pereira H., Faia R., Gomes L., Faria P., and Vale Z., 'Incentive-based and price-based demand response to prevent congestion in energy communities'. *Proceedings of the IEEE International Conference on Environment and Electrical Engineering and 2022 IEEE Industrial and Commercial Power Systems Europe, Prague, Czech Republic*, Jun 2022.

[35]   Mohanty S., Panda S., Parida S.M., *et al.* 'Demand side management of electric vehicles in smart grids: a survey on strategies, challenges, modeling, and optimization'. *Energy Reports.* 2022;8:12466–12490.

[36]   Melissa R., E-mobility: Electrifying the Way We Move, 2022. https://statzon.com.

[37]   Virta Global, The State of EV Charging Infrastructure in Europe by 2030, 2022. https://www.virta.global/blog/ev-charging-infrastructure-development-statistics.

[38]   Arcadis, Charging Infrastructure Market Report. An Analysis of Investment in the Electric Vehicle Charging Infrastructure Market, 2022. https://www.arcadis.com.

[39]   Efkarpidis N., De Rybel T., and Driesen D., 'Optimization control scheme utilizing small-scale distributed generators and OLTC distribution transformers'. *Sustainable Energy, Grids and Networks*. 2016;8:74–84.

[40]   Grigoras G. and Neagu B., 'Smart meter data-based three-stage algorithm to calculate power and energy losses in low voltage distribution networks'. *Energies*. 2019;12:3008.

[41]   Grigoras G., Noroc L., Chelaru E., *et al.*, 'Coordinated control of single-phase end-users for phase load balancing in active electric distribution networks'. *Mathematics*. 2021;9(21):2662.

[42]   Noroc L., Grigoras G., Dandea V., Chelaru E., and Neagu B., 'An efficient voltage control methodology in LV networks integrating PV prosumers using distribution transformers with OLTC'. In *Proceedings of the IEEE 20th International Power Electronics and Motion Control Conference*, Brasov, Romania, Sep 2022.

[43]   European Commission, Photovoltaic Geographical Information System – PVGIS, 2022. https://re.jrc.ec.europa.eu/pvg_tools/en/.

[44]   Rathod A.P.S., Mittal P., and Kumar B., 'Analysis of factors affecting the solar radiation received by any region'. In *Proceedings of the International Conference on Emerging Trends in Communication Technologies*, Dehradun, India, Nov 2016.

[45]   CENELEC, Voltage Characteristics of Electricity Supplied by Public Distribution Systems. https://standards.globalspec.com/std/14477184/pren-50160.

[46]   Qu R., Kendall G., and Pillay N. 'The general combinatorial optimization problem: towards automated algorithm design'. *IEEE Computational Intelligence Magazine*. 2020;15(2):14–23.

[47]   Pillay P. and Manyage M. 'Definitions of voltage unbalance'. *IEEE Power Engineering Review*. 2001;5:50–51.

[48]   Bollen M. 'Definitions of voltage unbalance'. *IEEE Power Engineering Review*. 2002;22(11):49–50.

[49]   Noroc L., Grigoras G., Chelaru E., Dandea V., and Neagu B. 'Voltage control strategy using the rule-based reasoning in LV distribution networks with PV penetration integrating OLTC-fitted transformer'. In *Proceedings of the International Conference on Electrical and Power Engineering*, Iaşi, Romania, Oct 2022.

Index